Engineering Manager's Handbook

An insider's guide to managing software development and engineering teams

Morgan Evans

BIRMINGHAM—MUMBAI

Engineering Manager's Handbook

Group Product Manager: Rohit Rajkumar

Publishing Product Manager: Himani Dewan

Book Project Manager: Sonam Pandey

Senior Editor: Rashi Dubey

Technical Editor: Joseph Aloocaran

Copy Editor: Safis Editing

Proofreader: Safis Editing

Indexer: Pratik Shirodkar

Production Designer: Nilesh Mohite

DevRel Marketing Coordinators: Nivedita Pandey and Namita Velgekar

First published: September 2023

Production reference: 1100823

Published by Packt Publishing Ltd.

Grosvenor House

11 St Paul's Square

Birmingham

B3 1RB, UK.

ISBN 978-1-80323-535-6

www.packtpub.com

This book is dedicated to the memory of my mentor, J Sam Frank.

– Morgan Evans

Contributors

About the author

Morgan Evans is an engineering management professional, mentor, advisor, and leader. She has worked in software development for the last 18 years as a coder, manager, and executive. She has led teams working on consumer products at scale in all sorts of circumstances, including multi-region and global teams across more than 30 countries. Morgan lives in New York City and runs a fractional CTO practice, where she lends her expertise to new ventures.

I want to thank Zebulon Holt,
who, through countless hours of mentoring, first taught me
what it means to be an engineering manager.

To Edward Cudahy, who taught me
how to lead world-class engineering teams.

And to the team at Packt for their help
and support throughout the writing process.

About the reviewers

Leslie Borrell is a technology and product executive with over 20 years of experience, using agile values, principles, and practices to deliver products and platforms for companies at all stages. She has demonstrated a unique talent for building engineering teams that overcome obstacles and challenges to deliver above expectations and find simple solutions to complex problems. She combines her deep understanding of agile values, engineering ethos, and practices that promote early delivery with her core values of empowerment and accountability to nurture inspired engineering teams. Leslie has always been at the forefront of the technology industry, seeking out the newest areas where she can apply her experience, tackle new challenges, and expand her thinking. As a change agent, she has an organization-wide impact, championing efforts such as reducing time to market, improving diversity and inclusivity, and enhancing quality and stability through continuous delivery practices. Throughout her journey, she remains passionate about efficiently scaling support and operations through intuitive internal tools, data, and tight feedback cycles. Leslie is utilizing her skills and experience to establish her own company called Carefully, a platform for the caring economy. The company's mission is to tackle a crucial global challenge - childcare.

Leslie's specialties include agile principles and practices, lean thinking, distributed development, project planning and management, analysis, continuous improvement, and continuous delivery.

Will Olson is an engineering manager with a knack for web application development. For over nine years, he's been deep in the coding trenches, and for the past three, he's also led distributed software engineering teams.

He got his start with an electrical engineering degree from the University of Nevada, Reno. Afterward, he began a career in which he juggled everything from working at a travel agency, building a custom CRM, to co-founding his own start-ups. Today, Will leads a software engineering team at Procore, a top construction management software company.

When he's not managing his team, you'll find Will exploring the indie hacker scene, where he continuously expands his skills and experiments with different emerging technologies.

Tristan Jung is an engineering leader with 10+ years of experience at LinkedIn and Twitter. Most recently, he led the core products engineering organization at Twitter and was the Twitter Toronto site lead. He led teams that built product features across the Home, Tweets, Profiles, Notifications, Search, and Trends surface areas.

Outside of software engineering, Tristan enjoys spending his time practicing photography, leading event organizations for various gaming leagues, and live production for events.

Table of Contents

Preface	**xv**

Part 1: The Case for Engineering Management

1

An Introduction to Engineering Management · 3

What are engineering managers responsible for?	**4**	Determine relative importance	10
		Fill in the gaps	11
Maintaining a team capable of serving business needs	4	Be a translator	12
		A checklist for building your schedule	13
Producing mechanisms to be self-sustaining and scalable	5	**How to prepare yourself for a career change**	**14**
Owning the reputation and impact of the team	6	Taking responsibility for the work product of others	14
Introducing the four activities of engineering managers	**7**	Saying goodbye to the rush of immediate results	15
How do engineering managers spend their day?	**8**	**Summary**	**15**
Start by asking	9	**Further reading**	**16**

2

Engineering Leadership Styles · 17

What is an engineering leadership style?	**17**	Philosophical origins	19
		Management theory origins	20
Leadership styles and their origins	**18**	**Engineering leadership style archetypes**	**21**
Natural origins	19	The commander and the servant	21

The coach and the delegator 22

The communicator and the co-creator 23

What is the right leadership style for me? 25

Summary 26

Further reading 27

3

Common Failure Modes for New Engineering Managers 29

Scenario 1—You don't know what is really going on 30

Scenario 2—You enable a narcissistic engineering culture 32

Scenario 3—You overshare information with your team 33

Scenario 4—You avoid making decisions 35

Scenario 5—You expect everyone to be the same 36

Scenario 6—You try to do everyone's job 37

Summary 38

Further reading 39

Part 2: Engineering

4

Leading Architecture 43

Setting the stage for good architecture 44

The environment 44

The building blocks 45

Gathering information 46

Decision methodology 46

Understanding the concerns of architecture 47

The breadth of concerns in architecture 47

Ownership and maintenance 49

Depending on open source 51

Naming things 51

Managing the architecture process 52

Building with a clear point of view 53

When you are or are not the architect 53

Emotion and other biases 53

Conway's law—part 1 55

Summary 56

Further reading 57

5

Project Planning and Delivery

Project Planning and Delivery — 59

Why do we need project planning?	60
Setting the stage for planning and delivery	60
The environment	61
Goal orientation	62
Project planning	63
Estimation	63
Prioritization	66
Assessing risks	67
Roadmapping	68
Forming the plan	69
Project delivery	71
Project kick-off	71
Good user stories	72
Removing friction	73
Project problems and solutions	76
You need to do more with less	76
You have scope creep	77
Summary	78
Further reading	79

6

Supporting Production Systems

Supporting Production Systems — 81

Creating a commitment to reliability	82
Ownership mindset	83
Taking pride in the work	84
Making a difference	85
Raising awareness of reliability	85
Metrics	86
Communicating metrics actively	86
Communicating metrics passively	87
Reliability solutions	87
Service objectives	87
Support documentation	88
Monitoring	90
Alerting	91
Service interruptions	91
Summary	93
Further reading	94

Part 3: Managing

7

Working Cross-Functionally

Working Cross-Functionally — 97

Demonstrating cross-functional leadership	98
Understanding your partners	98
Aligning with partners	100

Adopting a same-team attitude   100
Uniting team visions   100
Providing clarity on roles   101
Providing an aligned structure   101

Building strong relationships   **102**
Making yourself available   102
Further understanding partners   102
Helping partners understand you   103

Seeking and providing feedback   103

Difficult partnerships   **104**
Make your manager aware   104
Lean into the relationship   105
Work more defensively   105
If all else fails, escalate   106

Summary   **107**
Further reading   **107**

8

Communicating with Authority   109

Principles of communication   **110**
Setting expectations   110
Assuming the best   111
Saying no with care   112
Having an audience perspective   113
Maintaining authenticity   114
Giving feedback with radical candor   115

How to structure your communication   **116**
Format   116
Duration   117
Depth   117
Urgency   118

**Communicating with your
engineering team**   **118**
One-on-one meetings   119
Group communications   120
Personal commitments   121

**Communicating with your leadership
team**   **121**
Telling a story   121
Conveying broader value   122
Reducing uncertainty   123

Summary   **123**
Further reading   **124**

9

Assessing and Improving Team Performance   125

The classic stages of a team   **126**
Assessing engineering teams   **126**
Common pitfalls of assessing teams   127
Quantitative and qualitative measures   128
What is your success definition?   130
Assessing your team   131

Introducing team emergent states   **134**
Prioritizing desired team emergent states   135
Fostering team emergent states   136

Improving team performance   **137**
Motivating your team   137

Mentoring and coaching individuals on your team   139

Improving remote teams   141

Summary   141

Further reading   142

10

Fostering Accountability   143

Accountability and performance   144

Building an accountable team culture   145

Providing guidance   145

Promoting ideal behaviors   148

Accountability in practice   151

Summary   153

Further reading   154

11

Managing Risk   155

Why should you manage risks?   156

How do you manage risks?   156

Identifying risks   157

Prioritizing risks   159

Communicating and responding to risks   163

When and where should you manage risk?   164

Summary   165

Further reading   166

Part 4: Transitioning

12

Resilient Leadership   169

Introducing resilient teams   170

Why do resilient teams matter?   170

The engineering manager's role in creating resilient teams   171

Preparing your team for change   171

Managing yourself   171

Building a resilient culture   175

Building resilient habits   178

Preparing change for your team   180

Change communication   180

Change leadership   181

Summary   182

Further reading   183

13

Scaling Your Team   185

Recruiting and hiring	**186**	Giving structured performance objectives	196
Recruiting mindset	186	How to handle a bad hire	198
Writing the job profile	187	**Managing a large team**	**198**
Building your hiring team	188	Short-term and long-term planning	199
Interviewing practices	190	Delegating	199
Assessing candidates	193	Focusing on people over projects	200
Marketing your team	194	Preparing to switch contexts	200
Onboarding new hires	**195**	Making yourself available	200
Leveraging automation	195	**Summary**	**200**
Providing new-hire training	196	**Further reading**	**201**

14

Changing Priorities, Company Pivots, and Reorgs   203

Prioritization	**204**	Understand the changes	210
Methods of prioritization	204	Understand your team	210
Choosing a prioritization method	206	Be a guide and an advocate	211
Managing changes in priorities	**206**	Understand the directive	211
Prioritization context	207	Align the team and stakeholders	211
Prioritization dynamics	207	Build alliances	212
Prioritization solutions	208	Build momentum	212
User testing	209	**Summary**	**212**
Managing changes in objectives or structures	**210**	**Further reading**	**213**

Part 5: Long-Term Strategies

15

Retaining Talent   217

Why should you retain talent?	218
What does it take to retain talent?	218
Satisfaction with the work environment	219
Satisfaction with growth and opportunities	221
Satisfaction with their manager	223
Satisfaction with the company leadership	

and direction	225
Pitfalls of retaining talent	226
Can turnover be too low?	226
Can engineers be too satisfied?	226
Summary	227
Further reading	228

16

Team Design and More   229

Introducing engineering team design	229
Common team structures	230
Team characteristics	234
Conway's law—part 2	236
Lingering questions	236
What are squads, chapters, guilds, and tribes?	236

How many engineers can one person effectively manage?	237
What long-term goals should I have for my team?	238
What exactly is engineering culture?	239
What should I do if I disagree with my manager?	239
Summary	240
Further reading	241

Index	243
Other Books You May Enjoy	252

Preface

Hello and welcome to the *Engineering Manager's Handbook*! In this book, we use the term *engineering manager* as a catch-all term for position titles such as software engineering manager, software development manager, and web development manager. *Engineering manager* has become the prevailing way to refer to these positions within the world of digital product development, so for better or worse, we will uphold that convention here.

There are excellent books that cover leadership topics in general and engineering management in particular, but there are not many that attempt to introduce as many topics as possible, including available research papers. This book endeavors to be that resource. In true handbook fashion, this book provides breadth more than depth, with the intention of conveying to you the appropriate terminology and concepts to build upon as needed. In touching upon important topics for engineering managers, emphasis is placed on how to think about the problems of management in order to arrive at reasonable solutions. I hope to convey why behaviors are important rather than just describing the ideal behaviors.

Comprehensiveness is an elusive undertaking, so I welcome any feedback you may have on what might further complete this text.

Who this book is for

This book is for software development professionals who are currently in leadership roles or who aspire to be.

This book is intended to be useful not only to new engineering managers but also to those with experience. Since no two managers have identical career trajectories, there are almost always ideas that a manager hasn't been exposed to yet. In covering many topics, I hope to provide something new and useful to almost any engineering manager.

Additionally, this book is written to encompass strategies that apply to a wide set of workplace environments. Software development happens not only at tech companies but also companies of all sizes, growth stages, industries, and so on. Engineering managers who work within non-tech companies and non-product companies, such as those in corporate or small businesses, need strategies that work for them to provide good engineering management. This book is written specifically with these engineering managers in mind, providing differing approaches to consider under different circumstances.

What this book covers

Chapter 1, An Introduction to Engineering Management, poses the question, *"Why do we need engineering managers?"* and provides a rationale. It gives an overview of the obvious and not-so-obvious responsibilities of engineering managers. It provides foundational information on how engineering managers spend their time in different workplace contexts. Finally, it covers key concepts in the transition, from an individual contributor to a manager position.

Chapter 2, Engineering Leadership Styles, introduces what leadership styles are and where they come from. It reviews some of the most common leadership styles and how well they apply to different engineering team settings. It also describes how an engineering manager can examine and develop their own authentic leadership style.

Chapter 3, Common Failure Modes for New Engineering Managers, presents the common pitfalls and failure scenarios encountered by new engineering managers. You will learn why these failures occur and how they can be avoided.

Chapter 4, Leading Architecture, explains the engineering manager's role in technical systems design. It differentiates between the roles of manager and architect. It explains the responsibilities of the engineering manager and those of the architect, including what to do when they don't agree. Finally, it introduces Conway's Law and the importance of considering team design during the architectural process.

Chapter 5, Project Planning and Delivery, describes the engineering manager's role in the project and software delivery process. You will learn the key aspects of planning and delivering software, regardless of the project methodology used.

Chapter 6, Supporting Production Systems, presents the engineering manager's role in providing technology robustness. It describes how to build reliability into your team culture. You will learn common industry approaches to supporting and maintaining live systems and how to manage the moments when they inevitably fail.

Chapter 7, Working Cross-Functionally, details the best practices for working seamlessly with product management teams, design teams, and any other cross-functional partners, maximizing the productivity of this relationship. It also covers conflict resolution across functions and teaches you how to use RACI charts to ease the stress of collaboration.

Chapter 8, Communicating with Authority, introduces communication as a key area of responsibility for all engineering managers. This chapter argues that communication is one of the biggest force multipliers that engineering managers can master. You will learn best practices, how to structure communication, and how to communicate with specific audiences.

Chapter 9, Assessing and Improving Team Performance, covers how to evaluate the health and operations of engineering teams. You will learn techniques to optimize for success at the individual and team levels.

Chapter 10, Fostering Accountability, introduces accountability as a key characteristic of high-performing engineering teams. It explains in detail how an engineering manager can create a culture of accountability for their team.

Chapter 11, Managing Risk, explains what managing risk is and how it is a core responsibility and skill for engineering managers. You will learn how, when, and where to manage risks for your engineering team.

Chapter 12, Resilient Leadership, introduces the importance of resilience on engineering teams and explains the engineering manager's role in change management. You will learn why resilient teams perform better and how to instill a resilient culture in your team.

Chapter 13, Scaling Your Team, provides insider tips for scaling up an engineering team. You will learn about hiring best practices, techniques to onboard new hires, and how to manage a growing engineering team.

Chapter 14, Changing Priorities, Company Pivots, and Reorgs, answers the common questions of what to do if your organization has constantly changing priorities, unrealistic timelines, and a lack of focus. It details how engineering managers can lead with empathy during times of major change to improve outcomes for engineers and companies.

Chapter 15, Retaining Talent, walks you through a step-by-step plan to retain your engineering teams and create a great workplace environment.

Chapter 16, Team Design and More, presents basic concepts about structuring and operating engineering teams. You will learn the most common team alignments and the pros and cons of each. This chapter includes details on how individual characteristics affect team operations and how to consider Conway's Law when designing teams.

To get the most out of this book

This book assumes you have familiarity with foundational concepts in professional software development, such as unit testing and software version control. As such, software development terminology is often not defined so as not to be too tedious for the target audience.

Get in touch

Feedback from our readers is always welcome.

General feedback: If you have questions about any aspect of this book, email us at customercare@packtpub.com and mention the book title in the subject of your message.

Errata: Although we have taken every care to ensure the accuracy of our content, mistakes do happen. If you have found a mistake in this book, we would be grateful if you would report this to us. Please visit www.packtpub.com/support/errata and fill in the form.

Piracy: If you come across any illegal copies of our works in any form on the internet, we would be grateful if you would provide us with the location address or website name. Please contact us at copyright@packt.com with a link to the material.

If you are interested in becoming an author: If there is a topic that you have expertise in and you are interested in either writing or contributing to a book, please visit authors.packtpub.com.

Share Your Thoughts

Once you've read, we'd love to hear your thoughts! Scan the QR code below to go straight to the Amazon review page for this book and share your feedback.

https://packt.link/r/1803235357

Your review is important to us and the tech community and will help us make sure we're delivering excellent quality content.

Download a free PDF copy of this book

Thanks for purchasing this book!

Do you like to read on the go but are unable to carry your print books everywhere?

Is your eBook purchase not compatible with the device of your choice?

Don't worry, now with every Packt book you get a DRM-free PDF version of that book at no cost.

Read anywhere, any place, on any device. Search, copy, and paste code from your favorite technical books directly into your application.

The perks don't stop there, you can get exclusive access to discounts, newsletters, and great free content in your inbox daily

Follow these simple steps to get the benefits:

1. Scan the QR code or visit the link below

https://packt.link/free-ebook/9781803235356

2. Submit your proof of purchase

3. That's it! We'll send your free PDF and other benefits to your email directly

Part 1:
The Case for
Engineering Management

In this part, you will get an overview of what comprises the engineering manager role and why the work they do matters. You will learn about different approaches to engineering management and how to choose the right approach for your workplace. Finally, you will learn some common ways that engineering management can fail to produce the desired outcomes and how you can avoid those failures.

This part has the following chapters:

- *Chapter 1, An Introduction to Engineering Management*
- *Chapter 2, Engineering Leadership Styles*
- *Chapter 3, Common Failure Modes for New Engineering Managers*

1
An Introduction to Engineering Management

Software engineering and development teams are growing every year at an estimated rate of 25% (`https://www.bls.gov/ooh/computer-and-information-technology/software-developers.htm`). The digital transformation of businesses has led to steady increases in the number and variety of web and native app engineering positions. With the high cost of software engineers, all employers have a vested interest in the effectiveness of these teams. As software continues to consume the world, we have an increasing need for **engineering managers** to lead, inspire, support, and sustain our teams.

Many people have a hard time wrapping their brains around **engineering management** as its own discipline. When I tell others I am an engineering manager by trade, a common reply I get is, "*Oh, like a project manager?*" The skills of a project manager may be helpful to an engineering manager, but the work is different and requires a distinct skillset.

In this chapter, we'll introduce why we need engineering managers and what those managers must achieve in order to be successful. We will learn why the transition from engineer to manager can be such a difficult and jolting change, and we'll see why managing up is just as important to this role as managing down.

By the end of this chapter, you will understand which traits differentiate experienced engineering managers and what you must do to earn the respect and trust of your team.

This chapter is broken down into these main topics:

- What are engineering managers responsible for?
- Introducing the four activities of engineering managers
- How do engineering managers spend their day?
- How to prepare yourself for a career change

Let's dive in.

What are engineering managers responsible for?

An engineering manager's position exists to provide engineering teams with day-to-day leadership and representation. They keep alignment within the team and serve as the team's representative in cross-functional and leadership settings. In other words, engineering managers exist to produce long-term successful outcomes for engineers and companies.

When considering engineering manager responsibilities, you might think of a litany of tasks. Timely project delivery! Robust systems design! Mentoring! Hiring! The list goes on endlessly. The day-to-day tasks of an engineering manager are variable, but the broad responsibilities are the same: maintaining a team capable of serving the needs of the business, producing mechanisms to make the team self-sustaining and scalable, and owning the reputation and impact of the team.

This definition may leave you wondering: where are the fun parts? What about growing and teaching and building a strong engineering culture? While each of these things is important in its own right, an engineering manager should understand the order of operations: that these practices result from the necessity of producing good work, not the other way around. Your responsibility is not to produce good engineering culture; it is to produce good work that serves the needs of the business. Great engineering practices must result from the intention to produce great engineering work.

Let's go through each of these responsibilities in turn.

Maintaining a team capable of serving business needs

Engineering managers set their teams up for success by understanding the business and organizational settings they are in and preparing their engineers to thrive in those settings. As illustrated in *Figure 1.1*, engineering managers interpret a variety of inputs to determine how best to direct their team. This responsibility may require attention to many areas, such as ways of working, technology choices, staffing, team culture, and cross-functional interactions. While we may only know so much at a given time, it's helpful to consider anticipated future changes in needs as well as emerging technology trends:

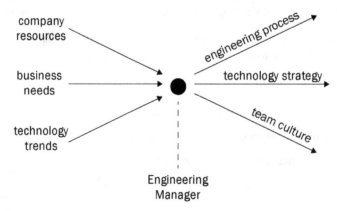

Figure 1.1 – Engineering manager inputs and outputs

Here are some questions you can use to consider this:

- Is my team working closely enough with business leaders to understand their domain and needs?
- Is my team capable of serving those business needs today or is it a stretch?
- Do we have the tools and processes we need to deliver?
- Do we have the talent we need on the team?
- Do we have the supporting cross-functional roles we need to do great engineering work?
- Do we have a shared vision to be inspired and make consistent choices?
- Do we have the support and guidance to grow and improve as engineers?

Not all of these considerations may be under your direct control to change, but they are still helpful to identify. It may be that you will need to spend more of your time influencing factors outside of the engineering team to help your team and company be successful.

Producing mechanisms to be self-sustaining and scalable

Engineering managers have a responsibility to optimize their teams. They improve engineering workflows and reduce dependencies and repetitive tasks. Self-sustaining teams minimize dependencies that hinder them in their efforts to achieve their objectives. Scalable teams minimize software delivery steps and eliminate bottlenecks. The mechanisms to achieve this may include the use of tools, conventions, documentation, processes, or abstract things such as values and principles. Any action that produces a tangible improvement in the speed, reliability, or robustness of your team's work is worth your consideration.

> **Theme – continuous improvement**
>
> In this chapter, we will introduce traits shared by experienced engineering managers that serve as recurring themes in this book. Great engineering managers are obsessed with continuous improvement for themselves and their engineering teams. They believe that there is always room for improvement and are open to new perspectives to that end.

Here are some questions you can use to consider how self-sustaining and scalable your team is today:

- What dependencies do we have on other teams?
- What dependencies do other teams have on us? Do we have clear contracts and interfaces that other teams can rely on?
- When we onboard a new engineer, how much time does that take from the existing engineering team?

- How much of our time is spent on manual interventions (such as responding to and resolving incidents)?

- Do we have the culture and values to be harmonious and productive?

- How do we respond to unexpected events? Are we resilient as an engineering team or are we brittle?

- Are roles and responsibilities clear for us and our cross-functional partners?

- Are we able to measure our work output and performance as a team?

- How many of our core processes rely on meetings?

- How many of our core processes have a **single point of failure** (SPOF)?

Answer these questions for yourself, honestly and without judgment. You will have plenty of time to formulate a plan to address areas for improvement as you settle into the role of engineering manager.

Owning the reputation and impact of the team

Engineering managers have a responsibility for their team's performance and output. They maintain awareness of the engineering team's commitments and progress. They act as a resource to help engineers do their best work. They cultivate sources of information outside of the team to have a balanced view of how their team is doing.

Here are some questions you can use to consider this:

- Am I currently aware of my team's reputation? Is it a strong reputation?

- Do engineers on other teams know what my team does?

- Do engineers on other teams like working with my team? Why or why not?

- Do people outside of engineering know what my team does?

- Do non-engineers like working with my team? Why or why not?

- Is my team impactful? Does it find ways to innovate and deliver creative solutions that produce a competitive advantage?

- Does company leadership understand the capabilities and value of my team?

- Is my team celebrated?

If you find that you don't know the answers to most of these questions, that is okay! You now have some talking points and information to gather for later.

So, now that we understand the general responsibilities of engineering managers, let's go through the activities and tasks that managers perform in service to these responsibilities.

Introducing the four activities of engineering managers

An engineering manager's work is made up of only four basic activities. Sounds easy, right? This is what they do:

- Engineering
- Managing
- Transitioning
- Big-picture thinking

How much of your time is spent on these activities will be determined by contextual and personal factors we will learn about later in this chapter. So, let's examine these.

Engineering—the planning and delivery of software for business needs—may be the bulk of your work as an engineering manager. You may contribute directly to these activities, coding alongside your individual contributors, but the primary role of an engineering manager in engineering activities is one of leadership, to guide and facilitate the best possible outcome and take accountability for decisions made. Let's break down the elements of engineering leadership activities:

- Leading architecture
- Project planning and delivery
- Supporting production systems

Managing—providing for the needs, well-being, and professional growth of your team—may be the bulk of your work as an engineering manager. This aspect of an engineering manager's role is incredibly impactful and can't be overemphasized in its ability to lead to great success or crushing failure from seemingly small changes. Let's go through the elements of management activities:

- Working cross-functionally
- Communicating with authority
- Assessing and improving team performance
- Fostering accountability
- Managing risk

Transitioning—guiding your team from one state of being to the next—is an inevitable eventuality for all engineering managers. Managers have a crucial role in preparing teams for change, contextualizing changes as they come, and providing teams with a sense of stability in an ever-changing world. Let's break down the key areas of transitioning:

- Resilient leadership
- Scaling your team
- Handling changing priorities, company pivots, and re-orgs

Big-picture thinking—is the time you devote to broader and more abstract questions of how to grow the utility and value of your engineering organization while retaining the progress and talent you have, touching on elements of all of the previous activities while thinking creatively to maximize impact and leverage. Key areas include:

- Retaining talent
- Team design

I have organized this book around these four activities as parts and chapters. If you are particularly interested in one topic, you can jump ahead, but it is recommended to read the topics in order for the best understanding.

Now that you understand what engineering managers are responsible for and the activities they perform in service to those responsibilities, let's go over how we might spend our day as engineering managers.

How do engineering managers spend their day?

No two engineering managers are likely to spend their time in the same way because the work is both contextual and personal. It is contextual because expectations can be vastly different in different settings. If you are a manager at a company with 10 engineers, it will be very different from being a manager at a company with 100 engineers, and that will be very different from being a manager at a company with 10,000 engineers. Similarly, your work will vary depending on your company's industry, leadership, growth stage, maturity, project, and supporting roles. These contextual factors combine in different ways to make every engineering manager's job unique.

An engineering manager's job is personal because individual skills and leanings vary. The technical background you bring to bear, your organizational skills, leadership style, personality, interests, influences, and beliefs will all play a role in the leader you become.

This means that successive managers of the same team are likely to have distinct approaches to their work, and your approach may change greatly from one position to the next. In this way, engineering management is as much an art as a science.

So, how do engineering managers know how to spend their day if there is no consistent approach? In this section, we will answer this question and teach you how to make these decisions for yourself like an experienced manager by going through the following steps:

- Start by asking
- Determine relative importance
- Fill in the gaps
- Be a translator

There may be many approaches to engineering management, but you should always start by asking!

Start by asking

As you start planning your days in a new position as an engineering manager, set yourself up for success by confirming the expectations of your leadership, peers, and team. Just as new engineers can deepen their understanding and progress more quickly by asking plenty of questions, curiosity can be even more critical to your success as an engineering manager.

Maybe you have a job description that clearly spells out the expectations of your position. If you don't have one, ask for it. Now is a great time to make sure you have read over this in its entirety and understand every point. Sit down with your manager and ask them the following questions:

- Should I take this job description literally or do you have another view?
- Can you clarify [points I have questions about]?
- I understand these items to mean [this]; would you agree with my assessment?
- In your view, what aspects of this are most important to my success in this position?
- Is there anything important to you that is not included here?
- How do you recommend I spend my day?
- Do you have any specific goals for me in my first week/month/six months?
- What do you see as the biggest challenges for my team to overcome?

Sometimes it can feel intimidating to drill your manager with questions, but they are there to help you be successful and will generally be happy to oblige. Set the stage by doing this in a one-on-one or by scheduling time specifically for a manager orientation session with them. You can say, *"I'd like to make sure I fully understand your expectations of me in this position, so I have a list of questions. I would appreciate it if we can go through these together."*

> **Theme – connection**
>
> Careful language is a powerful means of developing a connection with others. The ability to connect with those around you makes you relatable and helps others to view you with empathy and trust since they understand where you are coming from. Without sufficient time spent on connection, you may be distrusted, misunderstood, or viewed as out of touch. Connect with your engineers and those around you through your words, your writing, your reactions, and your sense of humor. Making strong connections involves a flow in both directions where you seek to understand others as much as you seek for them to understand you (see "Habit 5" from *The 7 Habits of Highly Effective People*: https://www.franklincovey.com/habit-5/). A mastery of connection allows you to influence those over which you have no real authority.

After touching base with your manager, it's a good idea to cast a wider net in your survey of expectations. Devote some time to inquiring about what your team would like from you day to day. Use the change in leadership as a reset to ask natural questions about what they enjoyed about previous managers and what they wished they had more or less of. Add in any questions about their expectations and history specific to your context. It's a good idea to make your intentions clear by saying directly, *"I'm coming up to speed here and I want to make sure I understand your expectations of me, any agreements you may have discussed with past managers, and any immediate frustrations or blockers you may have. I'm here to support you and I'd like to start by gaining an understanding of where you are today."* Be wary of making promises at this stage since you don't want to commit to anything you aren't 100% sure you can deliver.

If you work with product and design teams, set up some time with them individually to connect and start a dialogue by asking whether there is anything not outlined in your product responsibilities documentation that they are expecting from you. Ask similar questions on what they have appreciated about previous engineering managers and what they wished they had more or less of.

Plan for similar conversations with any other teams, partners, and stakeholders you expect to be working with. This could be analytics, quality or site reliability, infrastructure, business development, or many others. You may check in with your manager and ask who they think is worthwhile for you to connect with.

As you go through these expectation-surveying conversations, be particularly wary of commitments and overcommitments. Some may see a change in leadership and your enthusiasm as an opportunity to pile an unrealistic or self-serving wish list on you. Gather information, but resist the urge to appease everyone on day one. Aside from any serious HR-level misconduct, the right number of commitments to make at this stage is zero.

If your company has a strong memo culture, you may be able to save some time by sending out your questions in a written format. Even if you do that, it's helpful to do a bit of relationship-building and have a quick chat to ask any follow-up questions you may have.

Try to wrap up these conversations within a few weeks. By the time you finish your expectation survey, you will have a good idea of what everyone else expects from your time. Now, you can decide what you think about that. The purpose of gathering expectations is not so that you can blindly serve everything that everyone wants from you, but instead to raise your awareness. Your goal is to determine which expectations are duties you must fulfill and which need to be managed and adjusted.

Determine relative importance

So far in this section, we have learned how to survey others' expectations, and earlier in the chapter, we learned about the four activities we may divide our time across. Now, you can combine these sources to gain an idea of how you want to spend your day.

Take a look over your notes from the expectations survey you conducted. Looking at your notes through the lens of the four activities, does anything jump out immediately? Your manager may have had a clear emphasis on engineering, managing, transitioning, or some combination of these. Cross-functional peers often focus on engineering activities but sometimes highlight managing. Your engineering team is likely to focus on the management aspects of your role. Take note if the expectations across these groups fall along the same lines or are more spread out.

It's possible that the expectations are so spread out across the four activities that you are not sure where to start, or it may be that they seem focused on one area while you feel more passionate about another area altogether. How should you determine how to spend your time?

It's a good rule of thumb that your job is to follow your manager's lead, but your duty is to your engineering team. Start by following the advice you received from your manager and check in regularly that their expectations are being met. Leave yourself adequate time to serve the needs of your team. Keep in mind that you are a leader throughout this balancing act, and your sense of what is needed and good use of your time matters. Allow external factors to influence you while developing your own sense of time management.

Over time, you will develop more confidence and understanding of what the best uses of your time are. As you go through this process, keep in mind that since there are so many people depending on you, your time is precious! Be careful with it.

Let's go over a few more ways of determining how to spend your day.

Fill in the gaps

Another lens you can use to think about how to best spend your day is to think about your team's overall needs and output and where there are gaps. Some things by necessity can only be done by you. Other things can be done by anyone, but there may not be anyone available to do them.

If your team is understaffed for its workload, you should absolutely take it up with your leadership rather than overwork (more on this in *Chapter 5*). Aside from that, experienced engineering managers are typically expert gap fillers. They take advantage of their flexible time and flexible skillsets to move seamlessly through their day, lending free moments to whatever needs a little extra support to be successful. This might be building an internal tool for your team, designing a smoke test, teaching an analytics person how to pull a complex report, or helping a junior project manager. It could also be noticing the team is stressed and hand-delivering team members' favorite snacks with some encouraging words for a bit of relief.

Avoid the trap of filling gaps from a place of ego that *only I can write this piece of code* or *only I can do it in a short amount of time*. This type of thinking often leads new managers to sacrifice time that should be spent on leadership and management activities that only they can do, while depriving their team of a learning opportunity. You should be able to break down any work sufficiently and explain the right approach such that your engineers can tackle those challenges and grow their skills.

> **Theme – no ego**
>
> Engineering managers are there to guide teams, not dictate to them. This can be a mind shift for new engineering managers who may be used to doing everything their own way as individual contributors. We are all opinionated, but it is an important skill for engineering managers to be open to having their minds changed when new ideas come along. *Strong opinions, loosely held* is the rule of thumb. Not letting your ego get in the way of a change that makes sense is an essential trait of strong leadership. Over time, you should develop appreciation and respect for everyone's unique point of view.

As leaders of product development teams, engineering managers are uniquely positioned to help the entire team meet and exceed its goals by noticing gaps and lending a hand. Develop your own sense of where you are most needed and apply yourself there.

Now that we have learned a few immediate and practical ways to plan our days, let's move into more abstract territory.

Be a translator

For our purposes in this book, translating is the act of contextualizing information so that it can be understood by a particular group of people in the workplace. When you explain to product teams why it will take a long time to build a feature, you are translating. When you explain to an engineer why the company has made a change to project goals, you are translating. Anyone can be a good translator, but this is particularly relevant for engineering managers and is a key skill for impactful leadership.

So, how much of your day should be spent translating? The answer: more than you probably think. Translating is arguably the single most impactful thing an engineering manager can do with their time. The reason for this is the power of why.

Great articles, books, and talks have highlighted the power of why as a guide and motivator. Throughout your day, you may tell people what you want, what you expect, and what you recommend, but if the why isn't understandable to them or does not resonate with them, you have lost an opportunity to convince them. When you can translate the why completely, you put yourself in the best possible position for your desired outcome.

> **Theme – give work meaning**
>
> Great engineering managers find ways to give work meaning and make that meaning broadly understood. They align the realities of the engineering work they are tasked with to the aspirations and beliefs of their team members. For more on aligning difficult work with aspirations, see *Chapter 6*.

For your engineers, translating the why in a way they can understand and accept is a powerful tool for alignment and guiding decisions in the direction you want. You will spend less time answering questions, resolving disputes, and reworking mistakes if everyone is on the same page about the various whys from programming languages and tool choices to product priorities.

Translating outside of your team and upward to leadership (**managing up**) is oftentimes the most impactful translation of all. It can be a hard lesson to learn that the very best work by engineering teams can easily go overlooked or unappreciated if there is no one translating the value of that work in terms that leadership can understand. Ask yourself how widely your team's work is understood and who—if anyone—is doing that translating. Is there an opportunity for you to translate why your work matters to a wider audience? This is time well spent to gain awareness, adoption, and opportunity for your team's work.

So, where can you start filling the role of translator? In addition to your everyday conversations, you can be a translator in writing and in team rituals. Newsletters, documentation, memos, update meetings, or all-hands meetings are all excellent venues for a bit of translating. Use your judgment to determine the best audience, but don't shy away from translating your team's work broadly when there is an opportunity. You can approach your leadership and say, "*I believe the work we are doing on _____ is relevant to the entire company/department and I'd like to send out a monthly email update so that everyone can understand how this work can help them and the progress we are making. Here's an example I put together.*"

Spending part of your day as a translator pays dividends in all directions, helping others understand you, empathize with you, and become your supporters and friends.

A checklist for building your schedule

You can use this concise checklist as a guide when starting with a new engineering team:

1. Gather expectations from your manager, cross-functional partners, stakeholders, and engineers. Listen without making commitments.

2. Decide what is feasible and the best use of your time by blending expectations with your own judgment and intuition.

3. Estimate how much of your initial focus is needed on engineering, managing, transitioning, and big-picture thinking. Start to reserve time in your schedule/calendar.

4. Observe gaps that exist in your team's work and workflows and consider how you might fill those gaps for your team.

5. Make sure you are devoting some time to translating important contextual information between leadership, cross-functional partners, and engineers. Aim to consistently translate *why*.

6. Continuously adjust as new information becomes available.

Now that you have a sense of how you can plan your day, let's go over some specific reasons why taking on the responsibilities of an engineering manager can be a shock to the system and what we can do about it.

How to prepare yourself for a career change

It is important to set an expectation with yourself that you are going through a career change because of the magnitude of new skills and responsibilities you will be conquering. Engineers are accustomed to continuous learning, but the move from engineer to manager represents a move from finite and often well-defined goals (delivering features or owning a system) to what are more often complex and abstract goals such as keeping your team engaged or improving productivity.

This section focuses on preparing you to cross the chasm from engineer to manager without becoming burned out or overwhelmed by the weight of your new responsibilities. Let's learn how we can approach this career change.

Taking responsibility for the work product of others

Many engineers take great pride in their ability to solve everyday problems with code. Engineers build confidence by repeatedly endeavoring to build a solution and then delivering on that intention. The feeling of knowing exactly what you can and can't accomplish is a great one, but as you transition into engineering management, this comfortable feeling largely cannot come with you. When you lead a team, you will rarely have the same level of confidence because many factors will be out of your direct control.

It can be jarring and stressful to move from a world where you know all of your strengths, weaknesses, plans, intentions, and other commitments into a world where you have to try to ascertain those of several other people based on conversations and experiences. Even if you have worked with your team for long enough to know team members' abilities well, you cannot see inside their minds to know which competing or conflicting factors may throw a monkey wrench into your plans. This is inherently scary and will take some getting used to.

> **Theme – bravery**
>
> Engineering managers face discomfort and fear often. To be successful, engineering managers must conquer fears, believe in what they are doing, and be willing to stand up for what they believe in. Great engineering managers develop a sense of when they need to take a risk to make a difference, and they have the courage to act on their convictions.

The immediate and practical approach to the additional uncertainty of team leadership is to proceed with additional caution. As a new engineering manager, you may feel the urge to impress your peers and boss, but being overcommitted is a nerve-wracking experience that may have the opposite effect of what you intended.

When taking responsibility for the work of others, you also must prepare yourself to give up control of specific implementation details that do not ultimately matter. It will be stressful when your engineers do some things in a different way than you would yourself. It is now your job to provide useful constraints for your team to work within, such as conventions, security practices, performance rules, style rules, or code-complexity ratings. It will take some time for you to figure out how to best define and express these constraints, but what you don't want is to make your engineers feel like they have lost the ability to be creative and solve problems themselves.

Saying goodbye to the rush of immediate results

At some point in your first year or two of working as an engineering manager, you will hit a wall. Although the work of an engineering manager can be very fulfilling and satisfying, the feedback loop is much longer compared to working as an engineer. This is an adjustment that all engineering managers must go through.

As engineers and software developers, most of the time we get to tackle a problem, design a solution, see that solution merged and released, and enjoy the fruits of our work and a sense of accomplishment, all in a relatively short period of time. This short feedback loop becomes its own reward cycle, encouraging us to press on and tackle bigger challenges. We are able to push through frustrations and difficult moments because we know soon things will click and we will find the right solution.

As an engineering manager, you will have moments of immense personal satisfaction and a sense of accomplishment from helping and empowering your engineers. There will also be times when you are giving your full effort and you question whether it is making any difference at all. This can be incredibly draining at first. Quick and easy solutions are rare in the lives of engineering managers. Progress is usually gradual, and it can seem like a long time before you see the benefits of your efforts. Hang in there, be consistent, and you will eventually see results.

Summary

In this chapter, we introduced the purpose and goals of engineering managers. We learned about the broad responsibilities of engineering managers and the activities we perform in support of those responsibilities. Here's a recap of these:

- Maintaining teams capable of serving business needs
- Producing mechanisms to make engineering teams self-sustaining and scalable
- Owning the reputation and impact of their team
- Performing activities and tasks related to engineering, managing, transitioning, and big-picture thinking

We learned how engineering managers spend their days and presented steps for you to plan your own days, summarized here:

- Get a baseline of how to plan your day by surveying the expectations of your manager, peers, and engineering team.

- Focus expectation surveys on understanding what others expect from you and not on making early commitments you may struggle to uphold.

- Gain insight into the relative importance of work you might engage in by comparing the four activities of engineering managers with the notes from your expectation surveys.

- Observe what your engineering team needs and what members do and don't have time for to further determine how you can best contribute to the team's success.

- Avoid taking on hard technical challenges personally since they can absorb too much of your time and deprive your team of learning opportunities.

- Be a translator. Your unique understanding of engineering and the business setting puts you in a position to provide invaluable context for your engineers and your leadership team.

Lastly, we learned how taking responsibility for the work of others can be a stressful adjustment. Your goal is to find a balance between providing guidance and alignment without dictating or controlling. Progress as an engineering manager will feel much slower than progress as an engineer. Engineering managers must persevere without much feedback at times.

In the next chapter, *Chapter 2*, we will learn how to approach our responsibilities and activities with a style that fits the setting. We will introduce some of the most common leadership styles and how they apply to different engineering team situations.

Further reading

- Simon Sinek's *Start with Why* seminal TED talk (`https://www.ted.com/talks/simon_sinek_how_great_leaders_inspire_action`) and book (`https://simonsinek.com/books/start-with-why/`)

- Nancy Duarte's *Good Leadership Is About Communicating "Why"* (`https://hbr.org/2020/05/good-leadership-is-about-communicating-why`)

- Michael Jr's *Know Your Why* talk excerpt (`https://www.youtube.com/watch?v=LZe5y2D60YU`)

2
Engineering Leadership Styles

For engineering managers, **leadership style** has become an essential way of establishing and distinguishing who you are and which teams you are suitable to lead. Because of this, leadership style has become a frequent topic of interest in engineering management circles, making appearances in writings and talks. But *style* can seem like a vague term that doesn't convey a lot of meaning about what the concept includes or why it holds such a central role in the lives and livelihoods of engineering managers.

In this chapter, we will demystify what an engineering leadership style is and why it matters. We will review the origins of leadership styles and where they come from as relevant to modern software development. We will introduce common engineering leadership-style archetypes and discuss their effectiveness. We will close the chapter by learning how to develop your own style with intention and authenticity.

By the end of this chapter, you will have a firm grasp on what engineering leadership styles are and how to answer the question "*What is your leadership style?*" when it inevitably gets thrown your way.

This chapter is broken down into the following sections:

- What is an engineering leadership style?
- Leadership styles and their origins
- Engineering leadership style archetypes
- What is the right leadership style for me?

So, let's begin with what defines a leadership style.

What is an engineering leadership style?

Whether intentional or not, every engineering manager has a leadership style. This is because an engineering leadership style is no more than *salient beliefs reflected in actions over a period of time*. Let's break down this definition.

Leadership styles reflect salient beliefs. If your strongest belief is that engineering managers' role is to serve their teams, you will develop a servant leadership style. If your primary belief or interest is in mentoring and teaching your team, you will develop a coaching leadership style. If your primary belief or concern is controlling your team's actions, you will develop a commanding or micromanaging leadership style. Everyone holds many beliefs, some of which may be conflicting or competing, so it is the *salient beliefs* that tend to win out and define your leadership style.

Beliefs must be *reflected in actions* to become a part of your leadership style. Your actions develop and strengthen the abilities that support your leadership style. Practicing your beliefs cements them into skills and resources. If you believe one thing but act in a way that contradicts that, you may have some subconscious beliefs or drivers that are guiding you in a different direction. If you observe this happening to you, don't be discouraged. It is quite common for new engineering managers to need to work through some of their subconscious beliefs to better understand their own reactions and embody their ideals. An engineering manager's job is stressful, and it takes diligence and practice to act with intention.

Finally, beliefs must be reflected in actions *over a period of time* to become your leadership style. A leadership style is not defined by what you do once, twice, or occasionally. A leadership style is what you do consistently. It is what you demonstrate regardless of the circumstances or hardships.

Given this definition, you can develop your leadership style with intention by bringing awareness and active management to the beliefs you have about your role. Develop skills and resources to act on your beliefs consistently.

As a new engineering manager, give yourself time to develop your engineering leadership style. While you may hold strong beliefs that you work toward, you may initially lack the skills or experience to live up to those beliefs. This is an area where mentoring from a senior manager can help you gain perspective and progress.

Engineering leadership styles naturally evolve over time because beliefs and abilities evolve as you go through different experiences and receive new influences. You may also adjust your leadership style to different circumstances based on what you believe is appropriate for the setting.

Now that we have learned what leadership styles are and how to guide them with intention, let's take a step back to understand some of the origins of leadership styles.

Leadership styles and their origins

We have learned that on an individual level, leadership styles stem from beliefs and abilities, but what broader trends should we be aware of? Can the origins of leadership styles provide us with useful information today?

In this section, we will explore these questions and provide a brief overview of what we know about leadership styles from studies and writings. We will cover three foundational views of leadership styles and how you can incorporate the teachings from them into your efforts to develop an effective style for yourself.

We will briefly review the following:

- Leadership styles in nature
- Leadership styles in philosophy
- Leadership styles in management theory

Each of these views has something to teach us about leadership styles and why they exist. Let's start with leadership styles in nature.

Natural origins

Leadership is not unique to humanity; it can be seen across the majority of the animal kingdom. This means that leadership styles can be observed in all sorts of natural hierarchies, from ant colonies to migrating flocks of birds. When viewed collectively with what we know about human leadership and leadership styles, a few clear patterns emerge.

Nature shows us that the foundation of leadership is coordination (DOI: 10.1016/j.cub.2009.07.027). Leadership exists only where coordination is required. Not all species have leaders, just those that depend on groups for survival and so have a need for group coordination and social structures. With awareness of this, engineering managers can apply this knowledge to better understand the purpose and foundation of our roles. We can recognize that if our team is not well coordinated, our leadership style has fundamentally failed. Regardless of how it is achieved, an effective leadership style *must* optimize for group coordination.

Across nature, leaders tend to coordinate their groups in one of two ways: passively or actively. Passive leaders possess critical knowledge that group members retrieve from them individually as needed. Active leaders send signals of intentions to their group, with group members choosing to follow those signals or not. Passive leadership has the advantage of having less friction with less opportunity for dissent since there is no group forum to jointly push back on the leader. Active leadership has the advantage of being generally more efficient since direction is given broadly rather than individually. Since openness and efficiency fit well in most engineering cultures, the active style is often what we use in engineering team settings where leaders communicate direction broadly to their groups. There may be scenarios where a passive style could be appropriate to reduce friction or increase confidentiality on those teams.

Philosophical origins

One of the great arguments of history asks whether leadership should be rooted in personal charisma or in rational systems. These two broad leadership styles can be seen throughout history in many of the earliest writings on leadership from Sun Tzu to Plato and Aristotle (see *A history of leadership* in

The SAGE Handbook of Leadership, Keith Grint, SAGE Publications, 2011). The debate between these approaches and their moral and practical implications continues to be relevant today.

Charismatic leadership styles build loyal followers based on a cult of personality, setting a vision driven by emotion, and rhetorical skill. Rational leadership styles favor expertise, proven strategy, and systematic discourse, often taking a less visible role in their own administrations. Even in a technical setting such as engineering management, both of these styles can be widely practiced. Having a style in one of these directions doesn't necessarily mean you lack the abilities of or never practice the other—more that your dominant leadership style is characterized by those traits.

Charismatic leadership styles can have a mixed reception in a technical setting, but it is hard to argue with their effectiveness. On some level, most people want to be inspired, and charismatic leaders have the ability to connect with us through the deep emotions that drive subconscious decision processes in our brains (DOIs: `0.1257/089533005775196750; 10.1007/978-3-319-97082-0_3`). Charismatic styles can be a powerful driver of connection, which—as we learned in *Chapter 1*—is a key skill to master as an engineering manager. These styles also thrive in presenting the engineering team's work to executives and senior leadership.

The pitfalls of charismatic styles become apparent when leaders aren't able to follow through on their promises or deliver on the vision they have set. These styles must be able to follow through to be successful in an engineering setting.

Rational leadership styles can feel more expected in the logic-driven setting of software engineering teams and leaders. Engineers are accustomed to and appreciate explicitly proven methodologies and structured, rational discourse. Engineers value the depth of understanding provided by these styles, and during moments where goals are not attained, it can be easier for leaders to show how failures were systematic and not personal, retaining trust from the group.

On the other hand, rational leadership styles can struggle to attain the normative clarity of their charismatic counterparts. Not everything can be reduced to an obvious system or path forward, so these styles have a harder time gaining group consensus and acceptance of decisions. Removing uncertainty in choosing between strongly held, often emotionally driven preferences within the engineering team can be near impossible. Rational leadership styles must be able to cut through ambiguous or emotional situations to keep teams from becoming paralyzed with indecision or capitulating to the loudest voice in the room.

Management theory origins

When we think of modern engineering leadership styles, most of the concepts come from the world of business management theory and the prevailing wisdom there. Since the Industrial Revolution, researchers have studied and attempted to explain what defines good management practices. A simplified explanation of what researchers have learned is that over time, best practices in management have evolved based on protecting the sources of value creation in business (`https://hbr.org/2014/07/managements-three-eras-a-brief-history`).

When business value was driven primarily by the scale of owning massive factories and equipment, value creation centered on execution volume, and management practices were command-and-control with a view to produce-or-perish. As businesses continued to grow and become bogged down by their own size, value creation became centered on expertise in the systems, techniques, and processes that made production predictable and efficient, giving rise to management practices such as Six Sigma and supply chain optimization. This eventually gave way to the current era where business value is widely driven by knowledge work and value creation is centered on the workforce itself, so management practices have evolved to empathizing with, motivating, and retaining skilled workforces.

This reveals to us that as much as we may agree with current leadership styles that put focus on empathy and humanity, these styles have not come from a place of modern social expectations so much as the business realities and power balance of knowledge work. If the nature of our work changes in the future, we may find that expectations as to the purpose and methods of engineering managers change along with it.

Now that we understand some of the foundations of leadership styles, let's see how they have manifested into the modern leadership styles we know today.

Engineering leadership style archetypes

Most of us have had at least some exposure to the vast world of thought on modern engineering leadership styles and their best practices. There are seemingly endless sources presenting sets of leadership styles for engineering managers. It can be overwhelming to dig through this information. How can we approach these styles with intention and avoid just adopting whatever feels easiest?

In this section, we will learn what defines classic engineering leadership styles and in which circumstances they are most useful. You will learn how to look at these styles as a continuum instead of distinct personas. You will learn how to approach these styles as sets of practices that can be layered together to form a comprehensive style of your own, suitable to the needs of your engineering team.

To that aim, this section is broken down to compare and contrast leadership styles in the following subsections:

- The commander and the servant
- The coach and the delegator
- The communicator and the co-creator

Let's begin with two of the most widely discussed leadership styles: the commander and the servant.

The commander and the servant

The current zeitgeist has seen command-style leadership fall out of favor and servant leadership surge in popularity, despite neither being particularly new concepts. Commanders, sometimes called autocratic or dictatorial, are defined by the practice of setting a clear goal and directly telling their team how to reach it. For servants, on the other hand, instead of the team working to serve the leader, the leader works to serve the team. Commanders are more focused on the big picture while servants are more tuned to the individual needs of the team. As such, we can consider the commander and the servant as a continuum of leadership purpose, where at the one end we have driving alignment, and at the other end we have empowering individuals:

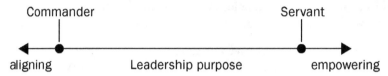

Figure 2.1 – Leadership styles on a continuum of purpose

A commander style is focused on driving alignment by taking a directive role. The advantage of this style is rapidly aligning a team. A commander style can be a powerful tool to get everyone on the same page during times of crisis or extreme change. When there is great uncertainty, this style can be appreciated for its ability to cut through the noise, give a clear path forward, and reassure followers with a strength of conviction. Within software development teams, this style is often criticized as disadvantageous for creative work and innovation since it doesn't leave much room for bottom-up perspectives or a proliferation of ideas. *This style needs to tread carefully when working with experts who have greater domain knowledge than the manager.*

A servant style is focused on empowering on an individual level through support and by creating opportunities. A servant style is a powerful means of building employee engagement in the creative work of software development. When team goals are widely understood and stable, this style is especially appreciated for its focus on engineer needs and well-being. Empathetic and low-ego leaders of this style inspire loyalty and commitment in their teams. On the other hand, this style has been criticized as reliant on the subjective expectations of followers. In other words, servant leadership can fail when those being served are too focused on individual desires. This style can lack the authoritativeness required in situations such as enacting an unpopular new policy. As engineering managers, not everything we do will be popular, and in those instances, servant leadership styles must find a way to gain team acceptance.

To consider adopting these leadership styles, you may ask yourself, *Do I need to rapidly align the team, or is my goal more to empower them as individuals?*

Now that we have examined leadership purpose, let's move on to guidance.

The coach and the delegator

While not as frequently discussed, it is likely that you have experienced at least one—if not both—of our next pair of leadership styles: the coach and the delegator. Coaches use mentoring and frequent feedback to drive growth and improvement in their teams. Delegators, sometimes referred to as laissez-faire leaders or self-organizing, simply route the work to the right persons and let them take it from there with limited input or interference. Coaches actively guide each member of the team while delegators let their engineers take the lead and set the course, so we can view the coach and the delegator as a continuum of guidance, from high guidance to low guidance:

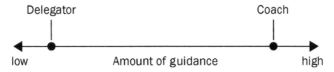

Figure 2.2 – Leadership styles on a continuum of guidance

A coaching style is focused on actively guiding each individual engineer. This style has the advantage of being very clear in expectations and performance feedback. Many engineers appreciate a coach's directness, which allows them to know where they stand. This style applies especially well to onboarding early career engineers, and any scenario where engineers might feel they have a lot to learn and would appreciate a guide to show them the way. At times, though, coaching doesn't land or isn't needed. If your engineer knows more about a subject than you do, coaching in that area can come across as patronizing or tone-deaf. Coaching succeeds when the coach possesses valuable experience that they can translate to the team. It fails when that experience is lacking or is not sufficiently conveyed.

A delegating style applies only the minimum guidance required to coordinate the team so that the engineer can determine the best way forward. At its best, this style puts the work in the right hands and gets out of the way. Skilled delegators give their teams the information they need to make good decisions and the freedom to come up with novel solutions and approaches. Many engineers appreciate the opportunity to experiment with their own ideas and take seriously the responsibility and trust they have been given by the leader. This style works particularly well in scenarios where engineers already have high expertise and motivation since it gives them the autonomy to flourish and demonstrate what they are capable of. Delegating styles fail when they do not provide sufficient information, when they are too distant and detached, or when engineers are not ready for autonomy in their work. Successful delegators must continue to take responsibility for their team's work and outcomes. They must not leave engineers feeling like they are out on a limb if things go wrong.

In considering these leadership styles for your role, ask yourself, *Does this engineer need more guidance or more autonomy in their work?* The answer may not be immediately obvious, so continue to ask this question over time.

Now that we have taken a close look at guidance, let's consider the source of vision.

The communicator and the co-creator

Lastly, we introduce the communicator and the co-creator. Communicators inspire teams to adhere to their vision by painting a compelling picture and convincing others to follow. Co-creators build networks of supporters and stakeholders by organizing and compromising. Communicators sell a singular vision, while co-creators organize a collective vision. In this way, we can think of the communicator and the co-creator as a continuum of vision, from a singular top-down vision at the one end to a networked multi-stakeholder vision at the other end:

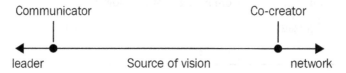

Figure 2.3 – Leadership styles on a continuum of vision

A communicator style is focused on selling an inspiring vision and mission for the team. This style benefits from a clarity of thought by circumventing potential *too many cooks in the kitchen* situations and driving a singular vision with a clear singular purpose. A singular vision frequently repeated by a communicator is much easier to follow and internalize across an engineering team. This style better avoids competing priorities or getting sidetracked and bogged down by too many ideas. Communicator styles help to simplify potentially complex projects such as building a new platform by keeping their singular vision top-of-mind at all times. To be successful, communicators must keep their visions grounded in the realities and capabilities of the team. This style fails when engineers lose faith that the vision is attainable or come to believe it does not reflect the realities of the work they are tasked with. When the singular vision no longer connects with reality, it will be ignored, and at worst, those leaders may be viewed as charlatans.

A co-creator style is focused on organizing a shared vision in service to a network of stakeholders. Co-creators assemble a vision by observing, listening, and evolving their ideas. Co-creators annex the goals of stakeholders into their vision to build a powerful network of contributors and supporters. This style can drive rapid scaling and innovation through high engagement and participation. Co-creators earn the respect of engineers and broader networks by working shoulder to shoulder with them to solve problems and being open to change along the way. This style is especially effective at building voluntary engagement and is common in volunteer open source project communities. Due to the multi-stakeholder nature, co-creators have to be careful about losing the focal point of their projects. Accepting more input takes more time, whether that input is used or not. To be successful, co-creators must find the right balance between accepting input and retaining focus. This style fails when projects lose all momentum, become bloated, or fail to please anyone by attempting to please everyone.

The ability to adopt these two styles is most often a function of the reality of your role at your workplace and whether you have the freedom to define or change the vision you are working with.

To recap, in this section, you learned six common engineering leadership styles and how to view them as ranges of practice to be applied in different scenarios. Take a moment to consider what your natural style might be in this context. Are you a serving coach? Or perhaps a delegating co-creator?

This section has shown us how common leadership styles aren't essentially good or bad but instead are tools to apply in different situations to give engineering teams what they most need to be successful. Work to master each of these archetypal styles so that you can meet the challenges of any situation.

So far in this chapter, we have learned about what engineering leadership styles are, how to cultivate our leadership style through examining beliefs, the origins of leadership styles, and what defines some classic leadership styles we may have heard of before. In the final section of this chapter, you will learn how to sort through all this information to craft your own authentic leadership style.

What is the right leadership style for me?

Back in *Chapter 1*, we learned how engineering managers spend their day and how this can be adjusted to the context of their position. That context can also be applied to your leadership style. An engineering leadership style is more personally driven than day planning, but context is still an important consideration. Your leadership style must be appropriate to your company culture, engineering culture, and team goals. You can have your own leadership style, but in cases where it conflicts with your company culture or engineering culture, it may become confusing for your team or unsustainable for you. There may be some cases where you can fly under the radar and break with company culture in service of creating a healthy atmosphere local to your team, but in practice, this can be difficult to maintain in the long term.

Contextualizing your leadership style to team goals means that you may need to lean into or away from some aspects of your natural style, depending on what you are working toward as a team. For example, leading your team toward a creative prototype will likely benefit from a different approach than the one you need to adopt when you're leading your team toward maximizing productivity while iterating on an existing system. Similarly, leading a team of highly opinionated senior engineers will likely need a different approach from leading a team of new associate engineers who've never worked together before. Factor in this team context and which goals you are working toward to help you fine-tune your leadership style.

Over time, feedback and results should also influence your leadership style. This is a read-the-room moment to grow your ability to connect with others. There will be times in your career as an engineering manager when you discover that you are being perceived in a way you didn't intend. Avoid becoming defensive and instead take the lesson you are presented with. Adapt your style to be more effective and more in line with your true intentions.

Periodically, it can be helpful to proactively gather feedback. One technique is to use surveys to allow anonymous responses since anonymity helps people feel more comfortable sharing concerns without fear of negative perception. Depending on your workplace, your department or HR may provide employee-survey tools to facilitate this feedback, so check with them before attempting to conduct your own.

If your workplace allows for it, try asking some simple, non-leading questions to start, such as the following:

- Is there anything you wish you had more of from your engineering manager?
- Is there anything you wish you had less of from your engineering manager?
- Is there any other feedback you would like to share with your engineering manager?

Along with feedback, results are another measure of your leadership style's effectiveness. Engineering team performance will be covered in depth in *Chapter 9*, but for now, you can ask yourself this: Is your team on a positive trajectory under your leadership? Are they trending toward improvement? What role might your leadership style have in the answer?

Now that you have learned how to incorporate workplace context, feedback, and results, you can start to craft your own unique leadership style. Follow your beliefs, develop supporting abilities, and course correct when appropriate. Keep in mind that the best leadership style is one that is an authentic reflection of who you are. Great leadership styles don't shy away from vulnerability. Being yourself and letting your personality shine through creates connections and builds trust with your team.

Summary

In this chapter, we learned what engineering leadership styles are, along with their foundations and historical context. We learned what defines leadership styles we commonly see in engineering teams, along with the benefits and drawbacks of those styles. Finally, we learned how to fine-tune our leadership styles over time.

We looked at the following key concepts:

- An engineering leadership style comes from salient beliefs reflected in actions over a period of time
- Leadership styles have a multitude of origins and are contingent on the industry practices and zeitgeist of the time
- Common engineering leadership styles today can be applied to suit varying levels of leadership purpose, guidance, and vision
- Your own leadership style can also be adjusted for workplace context, feedback, and results

Your engineering leadership style is a powerful tool for building consensus and connections with your team. Maintaining awareness and intentionality in your style will undeniably give you an advantage as you grow in your role as an engineering manager. I would challenge you to go through an interview process for an engineering manager position without being asked what your leadership style is. Use the techniques in this chapter to be well prepared when this occurs.

Now that we have learned about the role of engineering managers and their leadership styles, we will conclude *Part 1* of this book, with a careful examination of what *not* to do and why in *Chapter 3*.

Further reading

- *Management's Three Eras: A Brief History* by *Rita McGrath* (https://hbr.org/2014/07/managements-three-eras-a-brief-history)

- *Is Your Leadership Style Right for the Digital Age?* by *Jerry Wind* (https://knowledge.wharton.upenn.edu/article/the-right-leadership-style-for-the-digital-age/)

- *Finding the Right Balance — and Flexibility — in Your Leadership Style* by *Jordan, Wade,* and *Yokoi* (https://hbr.org/2022/01/finding-the-right-balance-and-flexibility-in-your-leadership-style)

- *Does Your Leadership Style Scare Your Employees?* by *Nihar Chhaya* (https://hbr.org/2019/07/are-your-employees-scared-of-you)

3
Common Failure Modes for New Engineering Managers

So far, in *Part 1* of this book, we have learned what engineering managers do, the different leadership styles that help them in their work, and themes: the set of traits embodied by the best managers. To complete *Part 1*, this chapter introduces common failures of engineering managers or, in other words, what not to do and why. Each of these touches on one or more topics that we will spend more time on in later chapters of this book.

In addition to knowing what to do, good engineering management entails knowing what *not* to do. The wrong actions on our part as managers can damage the trust placed in us by our teams. We might learn the hard way through painful failures, but occasionally, we are lucky enough to learn these lessons from the experiences of others. This chapter will prepare you for difficult scenarios that you might face and help you to recognize them before too much damage has been done.

We become engineering managers because we want to help other engineers, produce work we can be proud of, and protect our teams from pressures that can befall them. Sometimes, the best intentions can lead us to actions that end up undermining our own aims. In this chapter, we will learn about some of the ways that good intentions can lead us astray and produce the wrong outcomes. Then, we will talk about some of the common pitfalls that occur because of those intentions. Finally, we will learn how engineering managers can handle these situations differently and the techniques they can use to ensure better outcomes.

In software development, the term *failure modes* can be used to describe known failure scenarios. In this chapter, we will borrow the term and use it to describe ways in which engineering management can fail or produce undesired outcomes. Seeing how a system might fail gives us a better understanding of systems operations.

We will introduce the failure modes and their root causes in the following scenarios:

- Scenario 1—You don't know what is really going on

- Scenario 2—You enable a narcissistic engineering culture

- Scenario 3—You overshare information with your team

- Scenario 4—You avoid making decisions

- Scenario 5—You expect everyone to be the same

- Scenario 6—You try to do everyone's job

Each of these failure scenarios has something to teach us what engineering teams need and what happens when we are unable to provide it. Let's start with the first scenario, where you will learn what can happen when you don't invest in connecting with your engineering team.

Scenario 1—You don't know what is really going on

An engineering manager takes on the leadership of an engineering team. The new leader is quickly overwhelmed with the amount of work to be done, and they carefully go about their day reviewing project requirements, grooming a backlog, participating in planning ceremonies, reviewing code, running builds, and handling unexpected events and cross-functional needs. There are never enough hours in the day. The manager works to treat everyone with respect and fairness. The manager doesn't have as much time to connect with the team on a one-on-one basis as they would like but believes anyone would come to them with any serious concerns. Then, one day out of the blue, there's a resignation letter waiting in their inbox, and the manager is blindsided. This is a valued member of the team with years of institutional knowledge. It turns out the engineer was fed up with something happening on the team that the manager had no idea about, and that led them to consider and eventually accept another offer that came along. The manager thinks, *Why didn't they talk to me before it came to this?* Now, the manager must incur the considerable time and monetary cost of rehiring for this position.

This failure scenario can play out in many different ways, but generally, it involves the engineering manager being blindsided by the sudden revelation of something happening below the surface of their team. It might lead to resignations, productivity issues, or loss of morale in the engineering team. In whichever way it manifests, it's a highly stressful failure scenario for the engineering manager since it arrives suddenly and, often, urgently as a problem that they must now figure out what to do about. Often, managers feel responsible in this scenario because it could have been addressed before it grew into a bigger problem if only the manager had known about it.

You can never be completely protected from this scenario. Some amount of surprise will rear its head in your career as an engineering manager. What you can do is reduce the likelihood of it through deliberate action.

First, make time to connect with the engineers on your team regularly. There really is no substitute for connecting with people, building relationships, and giving them the opportunity to tell you what is on their minds. Meeting regularly gives valuable space for feedback in both directions.

Next, you can work to create a safe space where engineers feel comfortable sharing their concerns with you before they become bigger issues. To do this, develop your listening skills. Good listeners don't cut people off when they're speaking. They listen deeply and reflect what the speaker is saying back to them to be sure they understand. Use a simple reflection: "*It sounds like you are saying _____. Am I understanding you correctly?*" Then, avoid being overly judgmental or reactive about what they tell you. Receive feedback gracefully even if that feedback isn't delivered gracefully. Your responses in these situations will determine how comfortable your team will be in sharing sensitive information with you in the future. We will learn more about psychological safety in *Chapter 9*.

Finally, ask blameless questions to encourage your team members to be forthcoming. Sometimes, when questioned directly, a person who would not usually bring up an issue will mention it. A blameless question is phrased specifically to destigmatize problems and welcome honest responses; for example, saying, "*Do you have everything you need for your project?*" instead of "*Are you getting your work done?*" Here are a few questions to get you started, but add some of your own open-ended questions, too:

- *How do you like working with the rest of the team? How is it going on your project with [this person]?*
- *How are you feeling about your current project?*
- *You look like you have something on your mind. Care to share it with me? (If applicable)*
- *I noticed you've been quieter than usual lately. Is anything troubling you? (If applicable)*

Questions such as these send the subtle message that you value them, you see them, and you care if they are unhappy about something. Show an interest but avoid prying or being pushy—just let them know you are there if they want to talk.

Making time to connect, creating a safe space, and asking questions will not avoid all surprises, but it will put you in the best position possible to know what is going on with your team. Another thing you can do is be prepared for surprises before they happen. You can read *Chapter 11* for a detailed explanation of this.

In the next scenario, you will learn what can happen when you allow your engineering team to become too insular.

Scenario 2—You enable a narcissistic engineering culture

An engineering manager takes on the leadership of an engineering team. The team is skilled, effective, and efficient, and the manager is happy to lead such a capable team. The manager gets to work supporting and empowering the team. The engineering team is tight-knit; they enjoy each other's company and make challenging work fun with comradery and jokes. Sometimes, the manager notices that their team's jokes are at the expense of other teams within the company—for instance, the design team, the product team, the analytics team, or the business development team. Don't even get them started on the marketing team. The manager thinks this is harmless—after all, they aren't hurting anyone or being directly rude—so they ignore it and let the team have their fun. Gradually, the engineering team's lighthearted jokes evolve into cemented opinions. The design team is annoying, and the product team is incompetent. Everybody on the team knows that, right? Engineering is the only team that really knows what it is doing. This point of view starts to show up in some of the engineers' behaviors and actions. They ignore the people on the other team, assume the worst of them before checking, and don't bother to help them or advise them beyond what is absolutely necessary. The manager sees that their team is very capable, but projects aren't going smoothly. The work pipeline seems to take longer than it should and is prone to rework from misunderstandings between cross-functional peers. They seem to be missing opportunities to do impactful work since the teams aren't communicating closely as features are conceived. The manager realizes the work is suffering because the teams don't like working together anymore. They have allowed their team to believe that their part of the work is more important than the whole. Now, the manager has a long and difficult road ahead to fix this misalignment between the teams.

This scenario varies in terms of severity and how widespread the affected teams are, but it is a difficult failure to roll back. This is a case where an ounce of prevention is worth several pounds of cure. The engineering manager is in the difficult position of having to publicly change course and explain why they suddenly care about the relationships between the teams and no longer want these attitudes or jokes in the workplace. After taking this new position, the manager has an uphill battle to change the internalized beliefs that have been strengthened over time by the close bonds of the engineering team.

Fortunately, this scenario is relatively easy to avoid. It only requires consistency and firmness on the part of the engineering manager.

The first step toward avoiding this scenario is to embody the behavior you want to see from your team. Set the right example. Partner with your cross-functional peers from the teams that your team interacts with. Be in absolute lockstep with them as much as possible. Spend time with them. Eat lunch and have coffee with them. Talk about ideas together. Support each other publicly and visibly. Give them advice and candid feedback to help them whenever possible. Your team will see this and understand that this is what a good cross-functional relationship looks like.

Next, you must consistently instill that the product is the product; the code is not the product. It doesn't matter how elegant our systems designs are if they are in service to a lackluster or disjointed product experience. The result of the contributions of the cross-functional teams is more valuable than the sum of the parts. Whether an end-user-facing feature or a service interface, work should be considered from the user's point of view. In an effort across teams, one failure is a failure for everyone. We can't segment off the engineering team for a singular definition of success any more than we can say of a failed cake, "*but the flour is really good!*" You can avoid myopic definitions of success by consistently talking about the product rather than just the code base. Talk about the overall experience, usability, or developer interface. Make it clear that the end product is something the team should care about.

Lastly, have a no-tolerance policy on deriding other teams within your workgroup. Constructive feedback is great, but when it crosses the line into unhelpful, contemptuous, or divisive, speak up immediately. Rather than just silencing the remarks, take them head-on and create a teaching moment to correct behaviors and opinions. You can either use humor to rebuff the remark or flip the perspective on the situation. *Oh, the designer is being difficult? What's so difficult about it? If it's a hard design to implement, either you have the opportunity to do something new and cool (why not be excited about that?) or you have the opportunity to sit down and talk to them about what can be realistically accomplished in the time that you have available.* Encourage the team to partner with their cross-functional peers to elevate the work, not make them an adversary. You can read about more techniques for working cross-functionally in *Chapter 7*.

In the next section, you will learn what can happen when you become too relaxed about passing information along to your engineers.

Scenario 3—You overshare information with your team

An engineering manager takes on the leadership of an engineering team. Now, they have new responsibilities, a new peer group of other managers, and access to a new level of information. They regularly meet with the leadership team and hear news of upcoming work, new goals, and new directions on the horizon. The manager is told when to share and when not to share this information with their team. Some of these leadership ideas and plans come to fruition and some do not. Gradually, the manager becomes more comfortable with this abundance of information and its variability as it changes over time. They want to be an honest and transparent leader to their team, so they let some of this information leak here and there. The manager starts to tell their team more and more about what might be coming down the road, divulging potential policy changes, internal processes, possible projects, and product pivots that are being considered. The manager intends for this openness to build trust with their engineers and prepare them for the future. Yet, in some cases, the manager begins to notice that the information is not being received well. The engineers have a lot more questions than anticipated about what is happening, and the manager doesn't have any answers yet. Some of the engineers are confused and frustrated about not having clear answers or details. They become distracted by these new concerns and spend more time talking to each other about them. Some of the engineers are anxious that these changes might not be a good thing. Some feel a lack of stability and as though the winds are changing every week for no apparent reason. The manager realizes that their attempt to be honest has become a burden to the team, making them preoccupied with things that may or may not ever happen. Now, the manager is spending more of their time trying to calm down and console the team about hypotheticals that have not occurred. It is clear that morale and stability have both taken a hit, and the manager has a lot more work to do to walk back the effects of oversharing.

This scenario erodes trust and raises the day-to-day stress level of the team. Engineers might feel insecure about their work and the future of the company. Instead of thinking about engineering problems and solutions, they are spending time worrying about the unanswered questions they have and the general atmosphere that things are changing around them in unknown and uncertain ways. The engineers might start to disbelieve what the manager tells them since it seems to be variable and inconsistent to them. The manager now has to quell these vague fears and try to rebuild trust and stability in the team.

As ideas are discussed among leaders, engineering managers have exposure to sensitive and hypothetical information. This is both a privilege and a psychological burden since change is often a little bit scary for all of us. As engineering managers, we must bear the burden of change, supporting the ideas we agree with and arguing against the ones we don't. When we are too transparent about this process with our teams, we can very easily saddle our teams with unnecessary worries and make our own jobs more difficult. The reason we don't share everything we know with our teams is not to keep information from them but to shield them from the concerns of hypothetical change and give the leadership group more time to finalize and work through the details.

Avoiding this scenario requires restraint on the part of the engineering manager. Part of being a good translator of information is mastering timing. Find the balance between delivering information too early when it's disruptive and too late when the team might have heard it from elsewhere or missed an opportunity to give necessary feedback.

Deliver information to your engineering team with care and intention. As you become more comfortable with receiving confidential information and become closer to your engineering team, it is easy to fall into the trap of oversharing. Resist the urge. Giving your team advance notice should be rare to non-existent. One of the only times this might make sense is if there is a change being considered that could greatly impact your team, such as disallowing or switching out a software package that your team relies on heavily. In such a case, it might make sense to discuss the impact of this with one of your trusted senior engineers to get a second opinion on the implications of the change. Even so, you will want to mention to your manager beforehand that you are planning to discuss it with the relevant person on your team to raise any unconsidered issues.

Wait to communicate news to your team until you can do so from a position of clarity and certainty. Loop them in when you have the full story and necessary details. Prior to communicating, put yourself in your team's shoes and think through what questions they might have. Make sure you have prepared answers to those questions. For a more detailed explanation, refer to *Chapter 8* and *Chapter 14*.

In the next section, you will learn what might happen when you shy away from committing to decisions on your engineering team's work.

Scenario 4—You avoid making decisions

An engineering manager takes on the leadership of an engineering team. The manager wants to do a good job in their new position and develop a good track record. They hope to make the most of the opportunity they have been given, so they become focused on showing good performance in their role. They don't want to disappoint their manager or peers for any reason. As the engineering team goes about its work, the manager is careful not to overcommit. They make safe choices and follow practices that are well established within the company. When potentially impactful decisions come up, the manager defers the decision to the engineering team or outside of the team. If the decision can't be deferred, they look for a compromise of the available options so that no strong opinion is taken and all options are left open. As this pattern continues, the engineers on the team begin to consult the manager less and less often since the manager doesn't seem to want to be involved in those decisions, and they don't want to bother them. When certain decisions turn out to be the wrong approach, the manager can easily distance themselves from them when explaining the situation to others. Over time, the engineering team reaches a state where, for the most part, everyone is just doing what they want with little to no influence from the manager. Eventually, even when the manager does have an opinion, the engineering team doesn't seem to put much weight into it. The team resists the manager's direction and seems to be looking to other leaders inside or outside the team. The senior engineers question or antagonize the manager's requests. The engineering team is now running itself and treating the manager more like a PM. The manager realizes they have lost their authority over the team and that the team no longer looks to them for leadership. The manager is now spending their time trying to resolve power struggles within the team and trying to convince the team to listen to them.

Once it has advanced, this scenario is difficult to recover from because it centers on how the engineering manager is viewed by their team. Shying away from having opinions and making decisions demonstrates to the team a lack of ownership and a lack of accountability. Ownership and accountability are qualities we intrinsically expect from our leaders. We expect leaders to make decisions and own those decisions. Avoiding and deferring decisions erodes our own authority as leaders. When an engineering manager does not lead, the engineering team finds other leaders to fill the void.

To avoid this scenario, change your strategy from security by avoidance to accepting the risks of leadership.

First, be decisive. Decisiveness of opinion is a widely respected trait in all levels of software development. The ability to clearly state what your point of view is and, specifically, why you hold that point of view is a key skill in engineering and especially critical for engineering managers. You can absolutely factor in the perspectives of others, but a manager should be able to confidently state why a decision is right for their team.

Next, accept that being wrong sometimes is okay. Errors and mistakes are not the same as death. We make the best decisions we can with the information we have available at the time. If we never make any mistakes, we are probably playing it way too safe and incurring opportunity costs by never taking any risks. Being wrong sometimes is a part of the process of iterating. Being wrong sometimes helps us grow. This is valuable to demonstrate to your team and help them do their best work.

Lastly, become comfortable with accountability. Engineering managers are accountable—there is no getting around it. If you don't want to be accountable, this is not the job for you. As a manager, you will be held responsible for the successes and failures of your team, whether you like it or not. Your manager will respect you for taking responsibility directly rather than deferring the blame to your team. Your manager will respect you for stating what you could have done differently or how you could have been better prepared as a team. Your engineers will respect you for standing behind your decisions and not being afraid to take responsibility for the outcomes. You can read more about this in *Chapter 10*.

Next, you will learn the drawbacks of having an overly narrow definition of success for your team members.

Scenario 5—You expect everyone to be the same

An engineering manager takes on the leadership of an engineering team. The team seems to be really cohesive and working well together, consistently performing at or above expectations. The manager begins to assess their engineers' strengths and weaknesses. The manager wants to get an idea of the engineers' individual performance to mentor them and help them become even stronger as a team. The manager notices that a couple of the engineers are making the most code contributions and are really knowledgeable in key areas of the technology the team is working with—they are clearly the top performers. The manager is highly impressed with these top performers and believes their team can get even better if they can mentor the rest of the team to reach the level of these top performers. The manager starts taking action to get the whole team operating more like their top performers, talking to the rest of the team in one-on-one meetings about how to get their code contributions up and areas where they could become more knowledgeable. Try as the manager might, they can't seem to get the rest of the team working in the same way as their top performers. Gradually, the manager becomes frustrated that their plan is not showing results. The manager's frustration and somewhat obvious favoring of the top performers start becoming apparent to the engineers. The team that once was cohesive and happy starts to become a little less so. Some of the engineers lose confidence and start to think they are not doing a good job. Eventually, the work suffers as the team takes less joy in its work and doesn't see much point in going above and beyond what's required if it's not being appreciated. Now, the team's collective strong performance has become less reliable. They commit to less and seem to enjoy their work less. The manager's plan has backfired. Instead of raising the bar, the manager lowered the bar. Why didn't the manager's plan work? Now, the manager is busy trying to figure out how things went wrong and just get back to where they were when they started.

This scenario might not be the most severe, but it can be hard to discover and resolve. The emotional impact of the engineers losing confidence and feeling like they are not valued can be deep. The bigger danger is a loss of balance in the team. Engineering managers presented with this scenario could soon realize that the *less productive* engineers have been filling other critical roles that enable the team's overall success, such as communication, documentation, conflict resolution, or morale. You might find that the only reason certain engineers are able to contribute more code is that there is an unsung hero out there fielding all external questions that come to your team. There might be an engineer who spends considerable time with the product manager every week to make sure the handoffs are completely seamless or an average coder who is such a joy to be around that they singlehandedly make the work environment better and bring the team together. Usually, a strong team needs people playing different positions and playing them well. Engineering teams do more than just crank out code.

To avoid this failure scenario, resist the urge to use code contributions or any other singular metric as the primary measure of individual success. Evaluate the team's productivity as a whole. For individual assessments, look deeper to ascertain each engineer's true contributions.

Avoid being overly reductive in your assessments. Take the time to really observe the team and how they interact. Be open to the idea that a strong team is one where members complement each other rather than being identical to each other.

In your one-on-ones, try asking each engineer what they appreciate about the other engineers on the team. Ask them what the other engineers do that makes their day easier or better. Inquiring directly can be a great way to gain a deeper understanding of what engineers are contributing to the team that might not be immediately visible to you. Read *Chapter 9* for a more detailed explanation of how to assess your engineering team.

Finally, let's explore what may happen when you try to take up everyone else's slack on a project.

Scenario 6—You try to do everyone's job

An engineering manager takes on the leadership of an engineering team. The manager wants to do well in their new position and to help the team be successful in their work. They hope to make the most of the opportunity they have been given, so they have a view to do whatever is necessary to produce a good outcome. They are determined to make their first project a success, so they do their work meticulously. When problems arise with their cross-functional partners or engineers, the manager is quick to take up the slack however they can. The project is struggling, and needs arise that are outside of the manager's role and responsibilities, but they volunteer for them anyway. Someone needs to save the project, and the manager feels like they can handle the extra work. Before long, the manager's work day is filled with tasks fixing the work of others, and they are working nights and weekends to get their own work done. But the project is on track and everyone is praising their excellent progress, so they feel like they are on top of the world. The manager thinks, *I just need to get through this project and everything will go back to normal and I can have a sane schedule again.* But after the project is launched, there is a ton of post-launch work to do. The cross-functional partners and everyone else expect the manager to continue doing all the extra work they have been doing all along. The manager doesn't want to disappoint, so they continue to try to serve everyone's needs until they are completely burned out and exhausted. In the end, their work isn't as good as it used to be because they are always tired and can no longer sustain the same level of effort. The manager loses enthusiasm and becomes resentful of the work environment and angry at themselves.

This scenario is difficult to recover from because the manager has set expectations that they will perform an amount of work that eventually they cannot uphold. The manager gets a rush of pride when they are praised for going above and beyond, but this same pride makes it difficult to admit that they are in over their head and exhausted. It is hard to tell stakeholders and partners that you can't keep doing what you have been doing. It feels like a failure on your part, even though it shouldn't have been your responsibility in the first place. The manager must either face this mistake or reset by moving to another project.

Even if the manager were miraculously able to shoulder all the extra work indefinitely, they would face other problems such as boundary issues stemming from the appearance that anyone can dump their extra work on the manager.

To avoid this scenario, first, make sure that roles and responsibilities are defined. Ideally, you should have documentation outlining levels of ownership and contribution so that this is not up for debate. When project roles are clear, there is a basis for raising and resolving concerns that you might have. Read about defining project roles in *Chapter 5*.

Next, avoid this scenario by resisting the urge to save a project at any cost by way of assuming the responsibilities of others. Sometimes you may have the insight to know what is lacking in a project, and if that is the case, pass the information on to the right person instead of acting on it. Talk to your peers, manager, and leadership to give them the information necessary to get the project back on track. And if you have shared your concerns with everyone and nothing is done to address them, you may have to allow the project to fall short. While you certainly don't want the project to fail, doing unsustainable work only ensures that when it eventually fails, it drags you down with it. Read about when to allow things to fail in *Chapter 11*.

Summary

Throughout this chapter, we have seen how small oversights in your approach can lead to unexpected outcomes that are hard to undo after the fact. I hope this chapter will help you to avoid difficult and costly failure scenarios in your journey as an engineering manager.

Here is a review of the takeaways from these failure modes:

- Make time to connect with your engineers to help you avoid being blindsided by problems happening beneath the surface of your team. Leverage one-on-one conversations, create a safe environment for feedback, and ask blameless questions regularly.

- Take action to maintain relationships and mutual respect between cross-functional teams. Set a strong example with your own behavior, define success at the product level, and speak up immediately when you hear other teams being disparaged.

- Keep teams focused and happy by being careful about how and when information is shared. Resist the urge to divulge information before it has been finalized, prepare answers to anticipated questions from the team ahead of time, and deliver news with clarity and certainty.

- Maintain authority and leadership over your team by making decisions and owning the outcomes of those decisions. Internalize that it is okay to be wrong sometimes. What is more important is how we face accountability and recover from mistakes than that we never make any.

- Help team performance by taking a nuanced view of your engineers' contributions. Avoid being overly reductive in measuring the performance of each engineer and ask questions of each engineer to understand how they are helping each other to be successful as a team.

- Avoid overreaching in pursuit of project objectives since overextending yourself to save a project from failure will eventually bind you to that failure. Adhere to distinct project roles to keep workloads and cross-functional relationships equitable and balanced.

Understanding these failure modes and why they happen gives you the tools to change your approach before things have escalated. Familiarity with these scenarios makes them easier to recognize in your own team or the teams around you. Save time, energy, and frustration by keeping these lessons in mind.

With this, we have concluded *Part 1* of this book and gained a broad understanding of the role, work, and purpose of engineering managers. In *Part 2*, we will dive deeper into specific topics on how to lead engineering teams through their work on products and projects. For this, we will begin at the beginning and gain an understanding of how engineering managers lead architecture in *Chapter 4*.

Further reading

- *Mistakes I've Made as an Engineering Manager* (`https://css-tricks.com/mistakes-ive-made-as-an-engineering-manager/`)

- *The 7 biggest mistakes managers make* (`https://www.askamanager.org/2014/08/the-7-biggest-mistakes-managers-make.html`)

- *Some mistakes I made as a new manager* (`https://www.benkuhn.net/newmgr/`)

Part 2: Engineering

In this part, you will learn how engineering managers lead software development work. These chapters cover the most tangible aspects of engineering leadership. You will learn how engineering managers lead during architecture, project planning, project delivery, and production support.

This part has the following chapters:

- *Chapter 4, Leading Architecture*
- *Chapter 5, Project Planning and Delivery*
- *Chapter 6, Supporting Production Systems*

4
Leading Architecture

Software architecture is the process and product of creating technical **systems designs**. Architecture may include a specification of resources, patterns, conventions, and communication protocols, among other details. For the purposes of this chapter, we will also define architecture to include the selection of specific technologies.

Many great books have been written describing software architecture practices and principles, but this is not a software architecture book. Detailing the methods of software design is beyond the scope of this text. Instead, we will aim to describe the process and concerns of leading teams through software design phases.

Engineers love to solve problems and build systems. Most engineers would gladly build a system to solve just about any problem. It is the engineering manager's job to make sure that the team is using its time wisely and building the right system. In this chapter, we will learn how engineering managers can lead through the architecture stage of projects to produce good outcomes for their businesses and teams.

This chapter will move step by step through the preparation, design, and planning of systems. You will learn how to set the stage for architecture in different company and team environments. You will learn a broad set of concerns to consider during the architecture phase. You will learn strategies for when you are the project's software architect and when you are not. By the end of this chapter, you will understand how engineering managers can plan for systems to be impactful and maintainable.

The lessons contained in this chapter will help you determine the best approach to your architecture with the information you have available. To this aim, we will go through the following sections:

- Setting the stage for good architecture
- Understanding the concerns of architecture
- Managing the architecture process
- Conway's law part one

Let's begin by gaining a deeper understanding of the role of engineering managers during the systems design process and how they can set up the team for success.

Setting the stage for good architecture

Engineering managers have an interest in promoting good architecture because it produces better outcomes for their teams and businesses. The architecture process generates a plan of how to achieve product design goals. Complex work such as software development is more successful when it is planned because the planning process reveals problems early enough to solve and reconsider. An end-to-end plan helps to avoid rework and costs from errors in assumptions. It reduces future costs by paving the way for a code base that is less complex and more maintainable. It improves the user experience by creating an end product that is more scalable and performant. It also helps to scaffold the project delivery plan, since you have knowledge of which components need to be built in which order.

As an engineering manager, there will likely be many occasions when you are and are not the software architect for your project. We will dive deeper into the relationship between the two roles later in this chapter, but to start, let's assume you are not the primary architect for your project so that we may focus only on the responsibilities of the engineering manager.

Recall our *Fill in the gaps* section from *Chapter 1*. During architecture, the engineering manager observes the environment and available resources to fill in the gaps and ensure that regardless of what is needed from whom, the right actions are taken to produce the right systems design. As an owner of this new design, they must be fully invested in its success, whatever that may require.

As we have said, engineering managers work to produce the best possible outcomes for their teams and companies. In the architecture stages, this requires adapting your approach to the team's environment and ensuring you have sufficient information and processes in place to guide decisions.

The environment

To produce good work, engineering managers first ensure systems design is sufficiently conducted for their projects. Your role in this is often determined by the team environment. Consider the following questions:

- How accepted/expected is systems design in your organization?
- How formalized is systems design in your organization? Do many processes and practices exist for approving new architecture or not?
- How much latitude do you have in this design? That is, what degree of technical constraints or rules from your organization must you adhere to?

Engineering managers working in large organizations with formal processes and checks for systems design may have an uncomplicated role in architecture as compared to those working in organizations where those processes don't exist. Engineering managers in less formal organizations may find they have to do considerable work just to get time for systems design included in their project timelines at all. The table in *Figure 4.1* illustrates broad guidelines for the role of **engineering managers (EMs)** in differing **organizations (org)**:

	Low	High
Systems design acceptance in the org	EM ensures systems design is included in the project	EM ensures all information is available for systems design
Systems design formality in the org	EM defines the necessary steps to be taken (such as design deliverables, review steps, or an internal **Request for Comments (RFC)**[1] process)	EM ensures the required steps are followed
Systems design latitude on the project	EM ensures constraints are followed (such as technologies or specific services to be used)	EM creates and provides reasonable constraints
Systems design conflict and competition	EM ensures a sufficient set of design approaches is considered	EM ensures objectivity guides decisions over other factors

Figure 4.1 – Engineering manager focus in differing design environments

Based on your answers to the questions at the beginning of this section, you can determine what your role will be during the architecture phase. We see here that in low-formality, high-latitude environments, such as a start-up that has not established formal engineering practices yet, engineering managers can have a massive amount of work to do to guide teams. If this is the case for you, give yourself ample time for this important work and get as much support as you can from your senior engineers and product/ project managers. On the other hand, you might be in a high-formality, low-latitude architecture environment, such as a big tech company. In these contexts, architecture processes are more explicitly defined, formalized, and supported, so your focus can be on translating those expectations to your team and facilitating steps as needed.

Now that we have learned how to account for our environment, let's move on to how we can provide the information needed to encourage good decisions in the architecture process.

The building blocks

Systems architecture is the portion of any project over which the engineering team has the most control. This design is usually less of a collaboration between different functions and more clearly in the domain of engineers. As such, engineering managers have a high responsibility to own this process and its decisions. To produce the best decisions possible, you must have the right decision-building blocks: complete information to work from and a structured methodology to guide you. Let's go through each of these.

---

1   An RFC is a structured technical memo that announces an intention and requests that any interested individuals provide their feedback.

Gathering information

Well-built software can still completely miss the mark if those involved in its design didn't grasp the real problems being solved. It can be all too easy to focus on the wrong problem and end up with underwhelming results. Poor problem framing is incredibly common.

From *Judgment in Managerial Decision Making, Max H. Bazerman* and *Don A. Moore, Wiley*:

"Managers often err by (a) defining the problem in terms of a proposed solution, (b) missing a bigger problem, or (c) diagnosing the problem in terms of its symptoms. Your goal should be to solve the problem, not just eliminate its temporary symptoms."

Oftentimes, the requirements for a new feature or system come to the engineering team from a product manager or project manager. Whether or not this is true in your case, dig deep into the source of information for your system to suss out the real problems being solved. Problem exploration can be done by different roles, but the engineering manager or architect can gain a more nuanced view by conducting that exploration firsthand. Start with these problem-framing techniques and find more that work for your domain:

- Use stakeholder research; look at the problem from the perspective of internal stakeholders and end users, pairing with them or observing directly if you can.

- Use *why/how* laddering; think more abstractly by asking *why* and more concretely by asking *how* (`https://untools.co/abstraction-laddering`).

- Reframe the problem; write as many different ways you can think of to describe the problem in different terms.

- Challenge any assumptions; consider both *What if this assumption isn't (always) true?* and *Is there a perspective that may produce new options?*.

- Consider average versus extreme use cases; extreme use cases can highlight different ways of viewing a problem.

Keep in mind that in most cases, finding the right problem to solve is more impactful than finding a solution.

Decision methodology

The best systems design outcomes are the product of domain experts following rigorous decision processes. Engineering managers can facilitate optimal decisions by making sure there is a good decision process to work through. Following a structured process for decisions helps to avoid jumping to conclusions, missing a better solution that may be less obvious, or making decisions based on irrelevant factors.

Try this simple decision framework from *Bazerman and Moore, 2009*:

1. Define the problem
2. Identify the criteria (all relevant objectives for the system)
3. Assign weights to the criteria
4. Generate solution alternatives
5. Rate each alternative on each criterion
6. Compute the optimal decision

These steps are simple and easy to understand, but each represents a considerable amount of work. With effort and thoroughness given to each step of this process, you can ensure a valuable analysis and an informed decision. You may speed the process up by parallelizing some of the steps. While one person works on listing and weighting the criteria, others can begin generating solution alternatives.

Similar to a checklist, this approach produces consistency and a useful paper trail of justifications. Following a rigorous step-by-step structure such as this provides deep insights and ensures that decisions are made for the right reasons.

Now that we understand how to set the stage for the architecture process, let's move on to examining the concerns of architecture.

Understanding the concerns of architecture

Engineering managers work in vastly different work contexts where one of the starkest contrasts can be practices and expectations during systems design. Managers must be aware of their work context and have an understanding of the level of structure that is provided. They must fill in the gaps to ensure all considerations are covered to produce a successful design. To this end, this section will take a perspective of a low-structure work context in order to be broadly inclusive of what some engineering managers will need to provide to their teams. You may not need to worry about some of the concerns in this section, or they may be highly regulated and understood in your workplace—if so, feel free to jump ahead where desired.

To understand the concerns of architecture, we will learn about a broad set of factors that influence the viability of different architectural choices, dive deeper into some of the more intricate considerations around selecting technologies, and finish with a discussion of a closely related topic: naming systems.

Let's begin with a broad survey of concerns during systems design.

The breadth of concerns in architecture

When we think of systems design classically, we think of taking a set of requirements and capacity estimations and using them as the basis to compose a performant and reliable system. To produce the right design, engineering managers must also consider any and all factors that affect the ability to build, run, support, and maintain the design. Key concerns include cost (infrastructure, licensing, or otherwise), build staffing, testing strategy, deployment, post-launch maintenance, production support, security, scalability, intellectual property, and regulation. Design processes in your organization may or may not already address these, but the manager must make sure each is accounted for in the final design. Let's go through these briefly, as follows:

- The **cost** of running infrastructure is often considered in architectural design, but other costs and trade-offs are not always fully accounted for. The cost of building, running, supporting, and maintaining a system or feature can be compared to the cost of using a third-party service.

- **Build staffing** is usually accounted for in the product planning stages, but consider project risk and the likelihood that timelines run over.

- A **testing strategy** is not always considered in different company environments but can make a huge difference in the ongoing costs of a system.

- **Deployment** is not always considered during design phases. Starting with a robust delivery pipeline is a force multiplier for productivity.

- **Post-launch maintenance** is often overlooked in product planning processes at small-to-mid-sized companies. If teams will bear the increasing time cost of owning yet another system, the architecture may need to account for that.

- **Production support** can be overlooked in product planning at some companies. Consider how the system will be monitored and supported after it goes live to users.

- **Security** is usually considered in systems design. Decisions in this area are extremely impactful, especially if a system may be worked on by multiple or external engineering teams.

- **Scaling plans** may be considered or explicitly not considered at this stage (such as for early prototypes), but either way, this should be a clear decision widely understood by project/product managers and leadership.

- **Intellectual property** is not always considered in advance. Violations of ownership and attribution rights can have a massive financial impact.

- **Regulation** is a fast-changing area of rules on accessibility and data practices. Designs must account for rapidly changing and varied global regulatory environments.

Decisions made in the software design phase can greatly help or hinder each of these areas of concern. Engineering managers can make potentially massive impacts on business outcomes by incorporating these factors into the design phase and decision criteria. Which components you build or license, what data you collect or don't collect, and how you store and expose data are great examples of this.

Let's dive deeper into concerns of maintenance and support by considering how we can select technologies for our systems.

Ownership and maintenance

Engineers enjoy designing and building elegant systems. They select technologies that are the clearest fit with the best API and documentation. They may not always give thought to the consequences of ownership and maintenance, so it's the engineering manager's job to take a view of this. Use these three questions as different lenses to understand the trade-offs of design decisions and to inform your criteria for evaluating solutions:

- Is the purpose of this system/feature core to the business or not?

- Does my team have sole ownership of everything we build?

- How abundant is engineer time in the future?

Let's go through these one by one.

Is the purpose of this system/feature core to the business or not?

This question is intended to inform the technical vision of the design. In other words, what are the aspirations of this design? This is generally not a yes or no question, but a scale ranging from *This is our core business idea and technology* to *This just needs to work for the time being*. It is necessary to be explicit about where on this spectrum the project falls. It may be that a feature is very important to one or two stakeholders in the short term, but upon digging into it, they may just see it as a temporary stopgap. If your organization lacks the processes to capture this intent in the product definition, identifying this crucial information becomes essential for producing the right design.

Core business systems are investments. They aim to drive competitive advantage through differentiation. These features are where you want to focus your team's most ambitious efforts. On many teams, it is common to also need to work on non-core systems and features. These projects are still important, but the efforts you put behind them should be equal to their business impacts. Work on non-core systems should be designed and built to require minimal effort from the engineering team so that the team may devote the majority of its time to the core mission. Non-core efforts are the projects to license from third parties or use existing internal services when possible.

This idea may be obvious for something such as payment processing, for example, but keep it in mind for all non-core efforts, or you may find that an assumed-to-be-easy project can become a burden over time. Engineers or architects may view the initial design and implementation of a feature as something that can be done in an afternoon without realizing it will then evolve into its own workstream. A simple password authentication flow seems easy until further modifications and use cases are continually requested. A third-party service may be more time- and cost-effective even for small features. Teach your team to recognize and question these potential trade-offs.

Does my team have sole ownership of everything we build?

This question gets further into the maintenance and support implications of the design. In other words, is my team the sole owner to support and maintain this feature or is there a broader support system in place such as **site reliability engineering (SRE)**? The greater your team's responsibility, the less you may be able to accrue while continuing to support it all.

It may be that you are in a well-supported environment with shared responsibility. You may have customer support and technical writers to minimize the number of questions that come back to the engineering team. You may have site reliability engineers and infrastructure engineers to aid in robustness and produce supporting instrumentation. You may have developer advocates that free up engineer time by partnering with business development teams and clients. If your work context is fortunate enough to have many supporting roles, then ownership is a shared burden and the cost to the engineering team is lower. When the cost is lower, you can own more.

On the other hand, your workplace may be a lean environment where the engineering team does not have as many supporting roles. Your engineers and you may be the only people capable of performing work across a broad and demanding range of tasks. If you have sole ownership, then the cost to the engineering team is higher. When the cost is higher, you can own less and must be very careful about how quickly you are expanding your ownership.

How abundant is engineer time in the future?

Lastly, you can look to future expectations. You may or may not have access to detailed information about what to anticipate in the future, but this can be another perspective to consider for the design. Consider both the team growth phase and the roadmap. Growing your team's ownership footprint might make sense if your team size is also growing. Conversely, if your roadmap has committed your engineering team to many new builds in the coming months, you may want to pick and choose systems for which you can afford the ongoing cost of ownership.

Maybe your team and company are in a hypergrowth stage, rapidly scaling up to support your user base and product vision. In these cases, it may make sense to partner with your manager in planning your team's expansion and evolution of technology ownership. If you can count on continued growth, you may be able to take on more ownership for a greater product impact. If growth is uncertain, more caution is advantageous so that you don't end up owning more than you can manage.

Product roadmaps are another good source of predictive information. Dig into your roadmap to get a forecast of your team's future commitments and bandwidth. You may also have a technical roadmap that provides useful detail, or it can be very helpful to create one, outlining in advance which systems and features you anticipate owning and maintaining.

These questions can serve as valuable sources of information for determining which features it makes sense for your team to own the build and maintenance of. Now that you have determined what your team is capable of owning in a new design, let's look at how to approach the dependencies of that feature or system.

Depending on open source

While you and your team consider owning a system or feature, you will likely be considering its individual components and whether it makes sense to depend on open source packages. Much of this decision may come down to technical assessments of what is best for the architecture, but there are a few more managerial concerns. These may or may not have been considered in the technical assessment, but ensure that you have considered the component's maturity, supporting community, and license.

An open source component's maturity is often considered during the technical assessment. It's a good practice for engineering managers to make it clear to their teams what level of maturity is appropriate for dependencies since this can vary based on your company's level of risk tolerance. In some environments such as security, risk tolerance may be very low. It's a good idea to be highly explicit with your team about what the level of risk tolerance is and how that translates to the maturity of dependencies. If you aren't sure, definitely reach out to your manager or leadership team to find out in no uncertain terms, since this is not something you want to surprise you later. A policy for this can often be automated in your testing suite.

An open source component's supporting community is often considered in terms of size during the technical assessment, but it is not always considered in terms of source. The source of the community can also be impactful to the evolution of a component. For example, if most of the contributors to a project are from the same company, then that project will likely remain closely tied to the priorities of that company. This may or may not be a good idea to depend on based on your assessment of that company's future direction and how well it aligns with yours.

An open source component's license may or may not have been considered in the technical assessment, but as an engineering manager, you want to make sure this isn't overlooked. Depending on your work context, software licenses may be a common consideration during systems design. If they are not, it is a good idea to provide this level of structure and direction to your team.

These open source dependency considerations may be well established within your engineering team, but if not, you will want to do so as an engineering manager. Now that we have a sense of how to evaluate ownership and dependencies for our designs, let's go over some aspects of how we may name them.

Naming things

As you plan your design, you will likely start to think about naming. Naming things is often considered one of the few notoriously hard problems in software development since the implications of names are far-reaching and can be hard to change once established. Rather than looking at those classical problems of naming, we will briefly describe the challenge of naming, from an engineering management perspective: the creation of an internal lexicon of codenames within the organization.

Engineering teams often delight in giving systems and components fun or interesting names. Fun names are entertaining and amusing and can at times break up the monotony of the day. They add to the feeling that we are enjoying ourselves at work, but there are also drawbacks to creating these internal languages.

Having many codenames for systems can become a knowledge burden, an obscuring factor, and an unexpected learning curve that is difficult to roll back. In scenarios of rapid growth and hiring, they add another step to onboarding and training to learn a new language of what means what. Engineers spend mental cycles memorizing the codenames for services instead of reasoning about actual engineering problems.

You don't necessarily have to ban fun names altogether but consider using them sparingly to save effort down the line. Naming things as what they are is best for complete clarity, but there are different approaches and compromises that can be made. You can reserve these names for systems of a certain size, double name systems, or reserve fun names for releases/versions. Double naming means giving the system two names, for example, placing your deployment service at `deploy.myapp.com` but calling it *Deploy aka "Rocket."* Alternatively, you can give fun names to major releases or versions instead of the products and services themselves. This may be the best approach since names are clear and obvious, but you still get to have fun with the releases (or models or algorithms).

So, by now, you have learned how to examine a broad set of concerns, figure out what applies to your work context, and consider ownership, dependencies, and naming for your design. The factors you determine are relevant can be outlined as criteria in your decision process and appropriately weighted to inform your solution alternatives. So, let's go ahead and move on to techniques for guiding the solution process.

Managing the architecture process

The architecture solution process may be exciting, challenging, thought-provoking, or all of these. It is your job to get your team through it and to the right design. The next steps are to use the problem-framing information you gathered earlier to generate solutions, evaluate the solutions, and then work through the details of the chosen design approach.

At the beginning, your team may have many solutions to consider, but if your team is lacking in ideas, you can arrange for some brainstorming sessions. In brainstorming, there are no wrong answers and no ideas are too absurd. Brainstorming is most effective when focused only on idea generation, not editing. One seemingly absurd idea may inspire another one that can be a viable solution, so roll with the absurdity and encourage everyone to have fun with it. Embolden your team to think big and be creative during this process to maximize the ideas generated. From there, you can edit down into viable solutions and use your decision criteria to calculate the best approach.

Once you have determined a high-level approach, it's time to work on a detailed design. A few guidelines will help you with the design process.

Building with a clear point of view

Sometimes, project origination can take a long time. It may be rescoped, revised, or delayed. Regardless of the circumstances, on the architecture journey, the system needs to be designed with a clear point of view. The easiest way to ensure this is by having a single person in the role of architect with accountability for the systems design and technical vision. Ownership of the design does not mean making decisions unilaterally, and the person in this role should be open to input. But it is beneficial to have someone thinking holistically about the design. If ownership needs to be broadened or transferred, it can be very difficult not to lose some aspects in translation. Every effort should be made to document the design intention at various magnitudes and any relevant edge cases.

When you are or are not the architect

Throughout this chapter, we have focused on the responsibilities of engineering managers leading through the architecture stages. If you find yourself fulfilling both of the roles of engineering manager and software architect for your project, it helps to separate the two in your mind and in your duties. Give yourself time to think like an architect and separate time to think about the project as an engineering manager. Plan time to focus solely on systems design and arrange to be uninterrupted as much as possible. Create space in your mind for the possibility of having very different or even conflicting viewpoints from these two perspectives. Don't rush to converge on a solution too quickly.

On the other hand, if you are working with a software architect on the project, your aim is for a productive partnership. Make sure constraints are clear from the beginning to help the process go as smoothly as possible. Not only technical constraints, but clarity in the business context, project aspirations, and appropriate level of risk are foundational information to keep the two of you looking at the problem in the same way. It should go without saying, but always treat the architect with respect and kindness. Following the requirements, accepted constraints, and your rigorous decision process should preclude the vast majority of disagreements by keeping things objective. If you still manage to have a disagreement on the right approach, the next section on emotion and biases likely has your answer.

Emotion and other biases

A heuristic is a mental shortcut for easing the cognitive load of making decisions. Heuristics help us to make decisions in our day-to-day lives, but they can also lead us to wrong decisions or illogical conclusions. As engineering managers, an understanding of the most common heuristics can help us to avoid unconscious biases in our decision processes and recognize them in the behaviors and actions of others. To that aim, we will introduce the broad categories of heuristics that manifest as a wide range of biases so that we may have a general understanding, without enumerating the many individual biases. These heuristics are the availability heuristic, the representativeness heuristic, positive hypothesis testing, and the affect heuristic (Bazerman and Moore, 2009).

Availability heuristic

The availability heuristic states that people tend to overestimate what is most available in their memory, most vivid, or most emotional. For example, the most well-known stocks are typically considered by financial professionals to be the most over-valued. This heuristic can be helpful in quickly recognizing a pattern of failure that has reoccurred. It can also be harmful if you fail to consider other possibilities or jump to erroneous conclusions because of quickly judging from recent memory. This may surface in architecture as a bias toward recent or familiar design solutions.

Representativeness heuristic

The representativeness heuristic states that people tend to look for traits in individuals that correspond to previously formed stereotypes. We may unconsciously reduce others to anticipated archetypes or judge the potential of an idea based on how representative it is of successful or unsuccessful ideas we have encountered in the past. Some representative associations may be useful, but they can also be quite harmful and lead to prejudice and skewed decision-making. This may present in a design process as a bias against a pattern (*this approach never works*) or as a bias against the person from whom an idea originated (*this person never knows what they are talking about*).

Positive hypothesis testing

Positive hypothesis testing refers to the human tendency to intuitively examine how often a given case is true, rather than a more thorough examination of all outcomes. For example, if I were to ask you whether blond-haired people are great engineers, your brain would likely begin scanning for instances of people you know who are both blond and great engineers. A rational analysis would also need to equally consider people you know who are blond and not great engineers, those who are not blond and are great engineers, and those who are not blond and not great engineers. The intuitive tendency to only consider the positive case is misleading. Avoid this in your design process by confirming comparisons are systematic with positive and negative instances considered.

Affect heuristic

The affect heuristic states that in our judgments, emotional evaluation occurs before any reasoning takes place. These emotionally driven judgments might be accepted without further thought if a person is busy or under time constraints. This may show up in design processes where a preference cannot be explained or justified clearly.

If you believe one of these heuristics is at the root of a decision on your team, it is your responsibility to root it out. The way to do this is with probing questions. Since bias is not logical or objective, it cannot stand up to structured and rational questioning. It may take some time, but you will be able to uncover the issue. This is an area to be gentle and non-accusatory, since people may be very reluctant or have great difficulty in admitting where these ideas are coming from. Discuss it in private and with a sincere demeanor of wanting to understand; do not approach it as though you are concerned or that the person has done something wrong.

With that, your architecture should be converging on a rational, well-justified, and understood approach. Before we close out the chapter, we will briefly introduce Conway's law and how we can consider its effects before we move on to implementing the new design.

Conway's law—part 1

Melvin Conway is an American computer scientist who was prolific as a programmer during the 1960s and for decades beyond this. In 1968, he published a paper titled *How Do Committees Invent?* with the following thesis:

"Any organization that designs a system (defined broadly) will produce a design whose structure is a copy of the organization's communication structure."

In 1975, this thesis was referenced in the popular book *The Mythical Man-Month, Fred Brooks, Addison-Wesley*, in which Brooks dubbed it **Conway's law** (http://www.melconway.com/Home/Conways_Law.html). This adage has been referenced, studied, and heeded widely ever since. It is a frequent consideration for engineering managers since you most likely do not want your organization's structure to dictate the structure of your system. Having your systems design controlled by your organization's unique structuring and restructuring can be problematic when unintentional since your organization is often not a perfect representation of an ideal systems design.

As our architecture approaches finalization, we may ask what we can do with our design to account for the effects of Conway's law. There are several methods identified by researchers that engineering managers can use to address this, so let's go through them now (DOIs: 10.1109/52.795103; 10.1109/RESER.2013.14).

First, engineering managers can remove barriers to communication within and between teams. Communication is core to resolving the effects of Conway's law because the law is fundamentally a communications problem. Communication within a team is easier than communication between teams, leading teams to favor solving problems within their own team. Engineering managers can account for this by takings steps to facilitate better communication between teams. Make sure that both formal (meetings, ceremonies) and informal (chat) channels for communication exist and that engineers are comfortable using those channels. This is especially critical for remote and distributed teams. Collocation can be helpful but is not always possible or realistic.

Next, incorporate a modular design with clearly defined interfaces. Modular design is effective because it reduces the need for coordination. Define interfaces upfront with all stakeholders present. This is another area where relying on the information you gathered at the beginning of the design process is useful to bring to mind all use cases.

Lastly, make every effort to adhere to the interfaces and contracts you have planned. Some flexibility in the design process is beneficial, but frequent changes introduce too much additional coordination. Rely on patterns such as public/private methods to maintain consistency with original goals. Renegotiating interfaces carries a high cost and may compromise the fidelity of the design.

Now that we are familiar with Conway's law and how to account for it in our architecture, we are ready to wrap up and move on to our next stages of development. Later in the book, in *Chapter 16*, we will revisit Conway's law from the opposite perspective of how to account for its effects during engineering team design.

Summary

In this chapter, we have learned how engineering managers can guide and support their teams in architecting systems that not only cover basic requirements but provide insight and impact to the entire organization. Great architectural leadership has a profound effect on product outcomes and bottom-line business results. Remember these key takeaways:

- Begin setting the stage for architecture by understanding your environment and what you must provide to accommodate and support teams in that environment

- Finish setting the stage by getting your team their architecture building blocks: complete information to work from and a rigorous decision process to follow

- Take into account the broad areas of concern to be considered during the architecture phase and ensure they are included as criteria in your decision process

- Make sure project roles are defined so that your architecture can have a clear point of view

- If you are the project architect, delineate time for yourself to work from the perspective of multiple roles and points of view

- If you are not the project architect, provide the architect with clear constraints and context to work from and partner with them through the process

- Be aware of common decision heuristics and risks of bias, defusing them with deliberate inquiry

- Account for Conway's law with strong communications support, modular design, and clear consistent interfaces

With this, you should have a defined architecture for your project to use as the basis of your project development plan. Next, we will create that plan in *Chapter 5*.

Further reading

- *Are You Solving the Right Problems?* (`https://hbr.org/2017/01/are-you-solving-the-right-problems`)

- *How To Frame A Problem To Find The Right Solution* (`https://www.forbes.com/sites/palomacanterogomez/2019/04/10/how-to-frame-a-problem-to-find-the-right-solution/`)

- *A guide to problem framing* (`https://uxplanet.org/a-guide-to-problem-framing-ae58713364ec`)

- *Why How Laddering* (`https://wolfgangwopperer.com/notes/why-how-laddering`)

- *Judgment in Managerial Decision Making, Bazerman and Moore, Wiley, 2009*

- *How to Minimize Your Biases When Making Decisions* (`https://hbr.org/2012/09/how-to-minimize-your-biases-when`)

- *Three Ways To Avoid Bias In Decision-Making* (`https://www.forbes.com/sites/jarretjackson/2020/08/26/three-ways-to-avoid-bias-in-decision-making/`)

Project Planning and Delivery

The **planning and delivery** of projects is the most critical work we do in software engineering. Generally, all other work we do is in support of our ability to deliver working software. Some aspects of this process are under the control of engineering teams and some are not, but regardless, we have an interest in making the process go as smoothly as possible.

Planning and delivery practices may include specific project and product management methodologies, scheduling activities, team rituals, communication conventions, the creation of various software artifacts, and tools such as **version control systems** (**VCSs**). Engineering managers have a central role in negotiating these practices and helping them to work in the best interest of the engineering team and company.

This chapter will focus on project planning and delivery from an engineering management perspective. We will learn how engineering managers can approach planning and delivery across a variety of contexts and settings and how best to make a positive impact as a leader. We will introduce the core areas where managers can have an impact during planning and delivery. Finally, we will discuss common problems that arise and how we may resolve them. By the end of this chapter, you will be prepared to lead projects and respond to uncertainty in a variety of settings.

This chapter is organized into the following sections:

- Why do we need project planning?
- Setting the stage for planning and delivery
- Project planning
- Project delivery
- Problems and solutions

To begin, let's start by questioning why we need project planning in the first place.

Why do we need project planning?

If our goal as engineering managers is to deliver working software, not plans, we may question why we should bother with plans at all. Fundamentally, the purpose of planning is to reduce risk during project delivery. Without plans, we face significant risks of misalignment, miscommunication, and rework. Plans help us to avoid situations where team members have differing understandings of goals, constraints, and expectations. They help us to avoid wasted effort from misunderstanding what to deliver in what sequence.

Plans allow us to think through objectives beforehand in the hope of being prepared for delivery. Plans are useful when they preempt conflict, direct efforts in harmony, and align expectations. Plans are not useful when they waste valuable build time or provide a false sense of security, for example, by missing unknown unknowns.

Given the understanding that plans have useful features but are not foolproof, we can judge that they are worthwhile to some extent but shouldn't be overdeveloped or over-relied upon. In accordance with this, engineering managers can leverage plans to reduce risks to software delivery without becoming overly rigid or treating them as a panacea.

Timeboxing is a technique where you allot a specific amount of time to a task with the goal of doing the best job you can within that time frame. It helps to avoid spending excess time on a task inadvertently. This can be a good way to ensure you are not spending too much time on project planning (as well as some portions of delivery).

With this in mind, let's start by understanding what engineering managers can do to set the stage for planning and delivery.

Setting the stage for planning and delivery

In *Chapter 4*, we learned how engineering managers can assess what their work environment does and does not provide, fill in the gaps left by that, and set the stage for good architecture. Project planning and delivery are similar in that engineering managers can begin in the same way.

Depending on your work context, project planning may begin before, in parallel to, or after technical design work. The advantage to having technical design prior to project planning is that you can benefit from the discovery and information gathering already conducted. If planning occurs prior to technical discovery, only high-level planning can be done due to the lack of detailed systems knowledge.

To set the stage for planning and delivery, we will begin by assessing the environment and your role in that environment, as we did in the last chapter. From there, we will look at project goal orientation. Let's get started.

The environment

Engineering managers work across a wide range of workplace settings, from a highly structured and supported environment to a lean start-up. As you prepare for a new project, you may be partnered with a product manager, project manager, program manager, or none of these. Ask yourself these questions to assess your environment:

- Am I the primary planner of this project? If so, what support will I have?

- How formalized is project planning and delivery in my organization? What established processes or conventions exist that I need to be aware of?

- Is my organization flexible about planning, or is it rigid and exacting? Are they likely to be understanding of changes to the timeline or not?

You may find yourself working in an environment where you are partnered with a product manager for product direction, but instead of having a project manager, you are in charge of the planning and execution of the work. Many workplaces rely on engineering managers to fill the role of planner and day-to-day project lead during development. If you hold the responsibility for planning and tracking execution, you will likely have your hands full with these activities. Managers in this role need to focus on time management and follow a lightweight project methodology that allows them to balance project delivery with other duties as engineering managers.

On the other hand, if you are not in charge of the planning aspects of the project, that may make your day-to-day of focusing on engineering delivery a little easier. Your focus can instead be on making sure engineering needs are accounted for in the project process and on managing the relationship with the project manager. Taking the time to connect with the project manager and build a relationship will help them trust you and support you throughout the project.

Project planning and tracking may or may not be formalized in your organization. Either way, it is a good idea to assess and understand those needs at this stage, whether they be the use of specific planning tools, time tracking, or specific project delivery methodologies, such as Kanban. Make sure you understand what is required and of whom (if it is possible at this stage). If formality is low and you are working with the freedom to plan and track as you see fit, now is the time to consider what may work best for your team and the type of project you are doing.

Lastly, think about the flexibility of the plan before you start making commitments for yourself and your team. Develop a sense of what aspects of the coming project are flexible and what aspects are not. Talk directly to stakeholders individually about this, since they may not all see eye to eye. Your organization's *change averseness* in general and with regard to specific elements of the project is an important perspective to consider as you go through planning.

Seek clarity on your project responsibilities, methodologies, and level of flexibility before starting the project planning and delivery cycle. Understanding your environment and its expectations will give you confidence in your role in the project and help you determine your availability to take on project tasks.

Now that you are aware of the expectations of your workplace environment, let's consider what else may be useful prior to project planning.

Goal orientation

During upcoming project planning, you will be focused on the *how* of software development. How you will accomplish product goals, how you will meet timeline goals, how you can break down work into approachable pieces, and so on. Before you get to all of these *hows*, take another moment to consider the *whats* and *whys*. In between work on the architecture and the project planning is the ideal time for engineering managers to capture any goals specific to engineering priorities that were not outlined during the architecture phase. Engineering goals might include trialing new technologies, moving technology choices in a newly adopted direction, reducing tech debt by refactoring certain component(s), increasing the commoditization of a component, or many others.

Engineering managers look at every project as an opportunity to advance overall engineering goals such as these. Start assessing these opportunities by asking the following questions and continue with your own questions derived from your specific domain and engineering focus:

- Is there an opportunity to incorporate and trial a technology choice that may improve upon our current technology ecosystem? If so, is the level of risk of this trial acceptable and reasonable, in my opinion? What happens if the choice doesn't work out? What is my fallback plan?

- Is there an opportunity to advance migration of our technology ecosystem in a direction we have determined is our choice for the future, such as a particular pattern, programming language, or platform?

- Is there an opportunity to reduce tech debt by refactoring an underlying component as a relevant part of the project?

- Does it make sense to increase the commoditization of a service or component as a part of the project? Software commoditization refers to how generally applicable a system is or how well it fits into interoperable standards and patterns (`https://landley.net/writing/stuff/commodity.html`). For example, you might increase commoditization by adopting an adapter pattern to provide an interface that adheres to an existing standard.

You may have considered some of these goals during the architecture stage, but if not, now is the time to capture what engineering goals you would like to incorporate into the coming project plan. Some engineering goals may fit perfectly into the project needs, but some may require more consideration of the trade-offs in timeline and risk. Be prepared to discuss these trade-offs with the project team as you move into planning.

Now that you have set the stage with awareness of your environment and engineering goals, you can take an informed approach to your project planning.

Project planning

Project planning encompasses a wide range of activities in preparation for a project. Depending on your workplace customs and the project size, this may include scoping, budgeting, estimation, prioritization, assigning roles and responsibilities, creating timelines and roadmaps, setting milestones, and many other supporting tasks.

Regardless of what your responsibilities may be during the coming project, planning is the engineering manager's opportunity to set expectations with their project stakeholders, contributors, and themselves. Resolving expectations early on during planning helps to avoid misunderstandings and unpleasant surprises later when the project work is underway. Inward expectations to set for yourself and your engineers are **prioritization** and **roadmapping**. Outward expectations to set for the project contributors and stakeholders are **estimations** and **risks**. In this section, we will introduce these four elements of the planning process in sequential order and then bring them together to form the plan.

Estimation

Accurate estimation from a high level is hard. Unless you are replicating a project you have done before with known quantities, there will almost always be unknowns and confounding factors that conspire to prove estimates false. You may overestimate to provide yourself with a buffer in time for these unknowns, but this is not an ideal solution since you may end up encouraging your team to move more slowly than they are capable of (also known as **Parkinson's law**).

In *Chapter 4*, you began gathering requirements by understanding the problem you are solving. In addition to that information, there are two actions we can take to produce the most useful estimates possible. First, we determine the right approach to estimation, and then we develop the skill of breaking down work into small units that are easier to reason about.

How to approach estimation

In the previous section of this chapter, as we prepared for planning, we asked ourselves, *Is my organization flexible about planning, or is it rigid and exacting? Are they likely to be understanding of changes to the timeline or not?* Now we can apply this information to determine the best approach to estimation. You may be in an organization or product team that is more goal-focused than date-focused, where they are understanding and accepting that timelines may change along the way. In these settings, you can move through estimation more quickly and with more ambiguity so long as stakeholders understand and accept that estimates are high-level and timelines may be adjusted as new information is uncovered. On the other hand, you may be in an organization or on a project where delivery dates and timelines are rigid, such as when they are tied to real-world events or campaigns. In these more rigid settings, give estimates in such a way that you have thought through enough detail to have high confidence in your ability to deliver to the terms of that estimate.

In addition to understanding the level of detail we need to bring to our estimates, we can look at the direction of our estimates and what that means to the approach. In estimation, you are usually tasked with either estimating the time to accomplish features or the opposite scenario, where you are given an amount of time and asked what features you can deliver in that timeline.

In the first scenario, if there is some room for feature interpretation, it is useful to provide estimates as options and your recommendation. Your options may range from very minimal to more fully featured. For example, if the project is to implement a user-facing sign-in authentication flow, you might provide options that do and do not integrate with third-party identity providers such as *sign-in with Google*. Each option can have an estimate according to the level of effort of that implementation, such as 1 week to deliver signing in with email or 2 weeks to deliver signing in with email and signing in with Google. Complete this approach with a recommendation of which option you think would be best for the project and why, based on your knowledge of the project vision and goals.

On the other hand, if you are estimating what features can be accomplished in a given timeline, your estimates can focus more on the potential trade-offs you see. In this scenario, you might say, *In this time frame, we can deliver sign-in with email or sign-in with Google but not both*. Provide trade-offs and negotiate a solution that suits the project. This approach is useful for scoping projects in deadline-driven businesses.

Now that you have an idea of your approach and the level of detail needed, you can work on breaking down the project components.

Breaking down work

The key skill in producing estimates that are as accurate as possible is breaking down work into units that are easier to reason about. This skill is useful at the estimation stage and later during project delivery as you create and work through user stories or tickets. For this stage, you can start by outlining your project's deliverables in a nested tree of bullet points, going as deep as possible in listing the individual nodes that represent units of work for your project. Once you feel your project tree has the right level of detail, you can estimate the child nodes and sum those estimates for a rough total.

Here is a simple version of a project tree outline using our sign-in example:

- API methods:
  - Sign up
  - Sign in
  - Forgot password
- Frontend app views:
  - Sign up
  - Sign in
  - Forgot password

- Database:

  - Update user schema

  - Migration

- Test:

  - End to end

  - Load

In a real plan, you would need to include much more, but this is sufficient as an illustration of structure and depth. Along with these units of work, you can also consider what pre-development setup is needed and the time involved in that, and then what can be parallelized and what will have to be built in sequence. From this assembled information, you can get a sense of the effort and time required for your project.

Now that we have a sense of the appropriate level of detail, format, and units of our estimations, let's find out how we can increase the certainty of delivery by providing the assumptions that our estimates are based on.

Assumptions

An effort estimate is not complete without including its assumptions. Estimate assumptions include any and all underlying factors the estimate relies upon. Assumptions are especially important in more rigid estimation environments, but they are a good practice even where expectations are more flexible. Explicitly listing all assumptions helps to remove ambiguity and avoid misunderstandings during project delivery.

Your estimate assumptions may include the following:

- Technology dependencies such as access to specific services or resources, their SLAs and features, and/or how you may need to modify those resources. For example, *this estimate assumes we can use the existing SQL database for users and add new columns to its tables*. Or, *this estimate assumes we will depend on a third-party API for payment processing that offers deduplication and idempotency*.

- Scoping information, such as what features are in and out of scope. For example, *this estimate assumes we will develop not more than three application views including Sign Up, Sign In, and Forgot Password, using server-side form validation and no client-side validation and with no ability to view the user profile in this release*.

- Solution details such as what approach will be taken to fulfill the project requirements. For example, *this estimate assumes that we take a responsive approach to the web views using the same template design for desktop and mobile*.

- Personnel-based dependencies such as feedback or the access required for specific roles or individuals. For example, *this estimate for launch readiness assumes we will receive customer feedback within 3 days of providing the test link.*

Some of these examples may seem a little extreme, but I have included them to indicate the full range of information you may want to include in your assumptions in more rigid estimation environments, such as client contracting. These assumptions can act as the first level of risk assessment by raising awareness of the areas where the engineering team is depending on others to meet the project goals.

As you finish drafting your estimate, make sure you have incorporated feedback from subject-matter experts. It is a good idea to run your estimate by a couple of your engineering peers before committing to it.

Now that we understand estimates and assumptions, we can continue on to prioritization.

Prioritization

Prioritization is the process of ranking the features of a project by their importance. Most projects have elements that range from critically important down to nice-to-have but non-essential. Prioritization is a key area of importance to engineering managers because it is the source of most of the flexibility and adaptability in a project should you encounter unexpected challenges during delivery.

Many organizations will have a defined system for the prioritization of project sub-features, such as numbered levels: Priority 1, Priority 2, Priority 3…5. If this is the case in your organization, work with your PMs to understand the priority level of each item. If any of the project prioritizations seem extraneous to you, dig into why that item is a high priority to gain insight into the project drivers. Also make a mental note of what is low priority so that you can align efforts during the build with the overall importance of each feature and avoid investing undue amounts of time into work that isn't a high priority. If your organization doesn't have a defined system for prioritization, work with your stakeholders to introduce a priority ranking scale for the project.

If your stakeholders are resistant to prioritization and tell you everything is important, you can try time-based prioritization and effort-based prioritization. Ask your stakeholders what needs to be delivered first to gain a sense of the highest priorities on the project. Another way of phrasing this is to say, *What do we absolutely need for launch, and what could be a fast follow in a subsequent release?* You may also walk your stakeholders through the comparative effort of each unit of work to see whether that changes their minds on any prioritization. Non-technical stakeholders often do not have a sense of relative effort and may change their minds when they discover that a feature they wanted will take twice as long as the rest of the project.

Once you have gone through some of these prioritization activities, you will have a better sense of the goals of the project and where there is or is not flexibility in the delivery needs. Now that you have assessed priorities, you can consider potential risks.

Assessing risks

Project risks are any occurrences that could impact your ability to deliver the project as planned. Risks are important to communicate to stakeholders in any software project, and engineers are in an ideal position to anticipate them. At the project planning stage, focus on risk identification and communication.

Risk identification

Engineering managers get better at identifying project risks over time as they develop experience. Identifying risks early in the project process is extremely valuable since it gives you ample time to strategize and adjust plans as needed. The first step is to create a list of risks that you can keep track of and act upon when appropriate.

Start with the following sources to identify risks in your project:

- Any of your project assumptions that turn out to be false. Add the inverse of each estimate assumption you made as a project risk. For example, if you assumed the use of a third-party API, you could add a risk that the API isn't available or doesn't meet your requirements.

- Known unknowns are aspects of your project that involve uncertain efforts, solutions, technologies, or outcomes. These aspects are experimental in nature. Add them to your project risks. For example, *I believe we can find a solution to this problem within 2 weeks, but there is a risk that we may not.*

- Depending on your project, there may be hardware or networking risks.

- Staffing problems are a risk with varying levels of severity. Does your schedule include all public holidays and scheduled vacation time? Consider the impact if members of your team are out sick or resign during your project. Consider how likely you think this is. How much would your project be set back?

- Consider legal or regulatory risks. Are there aspects of your project that may require compliance with regional laws? Are there ways in which you might struggle with compliance or verification?

Another way to identify risks during project planning is to run a pre-mortem exercise. A **pre-mortem** is a session where you work backward and brainstorm how a project may fail. You can do this by yourself, but it is generally most effective with the whole engineering team or even all project contributors. One effective group approach to this exercise is to start with the hypothetical situation *this project has failed* and instruct each team member to write down on a sheet of paper why and how they think it might have failed. Then you can go around the room and discuss the possibilities. It may be very interesting to see what team members come up with and how similar or different the responses are.

Risk communication

Once you have completed your risk identification exercises, you can start the process of communicating the risks. If you are partnered with a project manager, this may be as simple as handing off the list of risks and allowing them to take over by further communicating, prioritizing, and coming up with remediation plans. If you don't have a project manager or anyone who is prepared to address the risks, it may be your responsibility to take the next steps.

If you expect to be handling the project risk management yourself, take a moment to plan your next steps. You will need to communicate the risks to stakeholders, but this may be information that they are not sure what to do with. If you have identified risks that are very urgent and impactful, raise them with stakeholders immediately, along with suggestions for addressing the risks. For the rest of the risks you have identified, you can put together a risk spreadsheet or matrix where you list them, prioritize them by impact, and come up with remediation plans/suggestions for as many as you can. Share this risk matrix with your project stakeholders and leadership. Use it to track and update risk actions during project delivery. If possible, create user stories or tickets to incorporate them into project preparation and delivery.

Now that you have an idea of project risks, you can go into roadmapping and release planning with a sense of where risks may affect timelines. See *Chapter 11* for a further explanation of managing risks.

Roadmapping

Depending on the conventions of your organization and size of your project, you may be creating a high-level roadmap, a more detailed schedule, or something in between, such as a collection of milestone dates. Regardless of which you are aiming for, this is a time for engineering managers to get a sense of the timeline and flow of events, capture what sequential needs exist, and identify any timeline-based risks.

During estimation, you broke down the project into units to be easier to reason about. Now, you will be thinking about those units across a more concrete timeline. This is the time to consider how your project methodology will integrate with the units you identified. If you are using Scrum, you might start to think about how this work could be distributed across sprints. If you are using Kanban, you might start to think about whether your units of work are of the right size and distinctness to be worked on independently.

One of the most important things about your timeline is that you capture the points at which there are sequential needs. The **Critical Path Method** provides a useful exercise where you enumerate the essential delivery milestones and calculate the sequence of activities to deliver the project (`https://asana.com/resources/critical-path-method`). This is important for avoiding time- and dependency-based blockers during your project. One of the worst feelings on a project is when your engineers can't make progress because they do not have what they need to continue. These might be sequential needs within the project, dependencies on items provided by other teams, or external dependencies on third-party deliveries. Capture the dates by which each of these is needed in your timeline. Where these sequential needs exist within the project, it is generally a good idea to deliver them as early as possible to avoid unnecessary risk downstream.

As you may have imagined while considering sequential needs, you may identify some of these sequential needs as timeline-based risks if you haven't already. Add these to your risk list or risk matrix.

At this point, you have captured the key pieces of information you need to establish shared expectations on your project, and you are ready to put them together into a project plan.

Forming the plan

Regardless of whether you are a contributor to or owner of the project plan, the most important thing to understand is that plans typically age very poorly. That is to say, plans change, and you should be prepared for that. Plan for how you can accommodate change and your plan will be much more useful. If you are forming the plan yourself or helping someone else, make sure you are change-ready and that you have some plan essentials we will introduce now.

Change readiness

Change readiness starts with forming the expectation of change and helping your team to have that expectation as well. While you will certainly aim to follow the plan, and the plan is what you would prefer to abide by, surprises happen, and in such cases, you should be prepared to be flexible. Convey this sentiment in your communications with the team.

Your previous prioritization work is your secret weapon for change-readiness. That list is where you will find wiggle room if you should end up needing it. Think about this in advance. What are the first trade-offs you might look to make? Where are there opportunities to adjust features to help delivery?

Stakeholder clarity

A good project plan needs to include the full set of stakeholders for the project. It is easy to capture the main stakeholders from business units or customer groups, but go a step further than this to identify fringe stakeholders in your project. Seek out representatives from the following areas:

- Security
- Compliance
- Legal
- Finance

Also, consider other engineering or infrastructure teams, quality teams, and data teams. Determine whether any of these teams may eventually be a gate to launch or a supplier of further requirements. Check with your manager for their ideas on other stakeholders to include.

You may also have shadow stakeholders appear on your project. For example, you may have a stakeholder partner for an entire project, and just at the end, their senior manager steps in and takes a different view of the work. Shadow stakeholders can be disruptive to a project, so try to anticipate where they may intercede and involve them at earlier intervals.

Project contributor roles

Projects go more smoothly when roles and responsibilities are clearly and unambiguously outlined during planning. Without ownership clarity, decisions tend to drag out longer than they need to while the team tries to sort through conflicting opinions. There are different methods of outlining roles and responsibilities that you can explore. One of the most common methods is the **RACI chart**. The chart lists each decision area and outlines who is responsible (does it), accountable (approves it), is consulted, and is informed.

Code	Name	Project Sponsor	Business Analyst	Project Manager	Technical Architect	Applications Development
Stage A	Manage Sales					
Stage B	Assess Job					
Stage C	Initiate Project					
C04	Security Governance (draft)	C	C	A	I	I
C10	Functional Requirements	A	R	I	C	I
C11	Business Acceptance Criteria	A	R	I	C	I
Stage D	Design Solution					

Figure 5.1 – RACI chart example (https://commons.wikimedia.org/wiki/File:RACI_Matrix.png)

Other methods you might consider for charting project decision ownership include the following:

- **RASCI (Responsible, Accountable, Supportive, Consulted, Informed)** adds a support role to include those who contribute directly to the project but are not the core responsible person

- **ARPA (Accountable, Responsible, Participant, Advisor)** attempts to avoid overreaching executives by limiting the decision powers of advisors

- **RAPID (Recommendation, Establish Agreement, Perform and Execute, Provide Input, Decide)** is more oriented toward decisions rather than tasks

These can be tailored to your work context and needs, but regardless of the format, what is most important is having thought through and agreed upon these contributor roles.

Communication strategy

A communication strategy for your project keeps your stakeholders, business leaders, and wider team informed and on the same page about your project. It saves time for you and your team from those seeking ad hoc updates or wanting to implement time-consuming tracking measures, such as individual hours reporting. The goal of your communication strategy is to be audience-focused. If you are responsible for the communication plan, write it from the perspective of each audience you need to keep updated, such as your stakeholder audience, your leadership audience, your customer or client audience, your engineering team audience, and your entire product development audience. For each audience you identify, determine the ideal communication method (format), frequency, depth, and duration. For example, an executive audience will likely appreciate a short, clear email update that highlights overall progress and challenges, while an engineering audience may have more appreciation for a more in-depth, longer write-up explaining technical problems and solutions. Be sure to consider more interactive and visual communication formats, such as demos.

By now, you have a plan with all the essentials: requirements, estimations, priorities, a timeline, risks, stakeholders, project roles, and communications. With that, you are ready to move into project delivery.

Project delivery

In project build phases, an engineering manager's focus is on maximizing the healthy productivity of their engineering team. In other words, you want to help the team move quickly while taking action to prevent burnout. In this section, we will learn these two aspects of project delivery: techniques for moving quickly and techniques for avoiding burnout. These goals can share the same solutions or complement each other, so they are interleaved through the following delivery topics: project kick-off, good user stories, and removing friction.

Project kick-off

The project kick-off is how you initiate your project. It is often a meeting or a series of activities to announce and begin work on a new project. For engineering managers, project kick-off is an opportunity to set up the project for success by aligning the team and creating some excitement about the goals.

Whether or not you are tasked with organizing the kick-off activities yourself, your aim is to ensure there is a project vision set for the team. A project vision includes the purpose of the project, how it ties into the mission of the product or company, and any additional contextual information that helps the team understand the nuances and importance of the work you are doing. The project vision serves many purposes. It aligns the team with what you are trying to accomplish and how you plan to accomplish it. It helps the team move in the same direction and avoid confusion. It provides contextual information that helps them make decisions that are in line with the goals of the project. It bonds the team together as collaborators working on a shared vision. It gives them a sense of purpose in their work and that what they are creating is important and impactful to the world around them. It generates enthusiasm and belief in the work they are doing.

Giving work meaning is a foundational technique for avoiding burnout. People with a sense of purpose in the work they are doing are more resilient and better able to persevere through challenges. It doesn't have to be a world-changing purpose; it can be something as simple as making other engineers' lives a little better. If they believe in the work that they are doing and its impact, they will be better prepared to overcome difficulties along the way.

Set your project vision by explaining it in a meeting and capturing it front and center in your project onboarding materials (such as documentation, a dashboard, or a landing page in your issue tracker). Continuously refer back to it to keep the vision top of mind as you move through project delivery milestones.

Good user stories

User stories are a common method of conveying the requirements of a unit of development work by describing what an end user should be able to do in a system. Classic user stories typically follow the format: *As a [type of user], I want to [what], in order to [why]*. To assess the quality of user stories, Bill Wake created the INVEST criteria. These criteria include the following story characteristics:

- Independent – being self-contained
- Negotiable – leaving some room for creativity in implementation
- Valuable – having a clear benefit
- Estimatable – able to resolve an effort estimate
- Small – of a manageable size
- Testable – making test development possible

Whether you are creating user stories, tickets, issues, or any other format, there are a few critical elements that help your team understand how to approach a problem and get the solution right the first time. These include making the stories an approachable size, defining acceptance criteria, and giving framing information such as priority and benefit.

Story size

During project planning, we learned how to break a project down into units that are easier to reason about. This skill is useful not only for estimation but also for dividing up the work among the engineers who are building the new functionality. Depending on the size of your project and the detail level of your estimations, the units you defined previously may be the right size, or they may need to be broken down further. Capturing the work in small enough units removes ambiguity from the features and allows engineers to think about a simple problem rather than a more complex piece of composition. Small units of work are more approachable, inspire confidence, and give a sense of accomplishment as engineers work through them rapidly.

Acceptance criteria

Acceptance criteria are a list of conditions that must be met for a user story or ticket to be accepted as complete. In other words, it is the definition of done. Good and thorough acceptance criteria in your requirements help your engineers move quickly by understanding how to approach a problem and by avoiding misunderstandings, revisions, and rework. Acceptance criteria should include input from stakeholders, cross-functional partners, and any technology constraints. Make sure to capture the input of the fringe stakeholders you identified in planning, such as legal and compliance needs. It can be demoralizing and discouraging for engineers to have to continuously go back and revise work because important criteria were not identified ahead of implementation.

Framing information

A sense of relative importance and/or benefit of the story is the final piece of useful information to give engineers the context to make good decisions. Along with having a digestible size and clear criteria, framing information outlines how they may want to think about the problem they are solving. Relative importance can be drawn from the prioritization exercise you performed earlier. A benefit statement further describes the purpose of the story and adds nuance to the requirements so that engineers can think creatively. For example, if your story is *As a user, I want to be able to access the search input from any page of the app so that I can quickly find what I'm looking for*, then your benefit statement might be that this feature is intended to save the user time and reduce frustration. This framing information provides context as to the point of view of the end users and stakeholders, allowing the engineers to empathize and consider the problem from different perspectives.

Now that we know how to capture requirements, let's move on to removing friction from the delivery process.

Removing friction

Once you have the vision and requirements, your next area of focus as an engineering manager is identifying and removing obstacles to rapid delivery. There are many potential sources of obstacles and friction. These don't just slow down your project; they also drain the energy of your engineering team, create stress, and increase the likelihood of burnout. Removing friction does double duty to speed up delivery and improve morale. The main areas of friction to address are processes, people, and tooling.

Removing process friction

Pain points from project processes are incredibly common. Most engineers have, at some point, struggled with project processes interfering with their day and productivity. This friction may be from meetings that are excessively frequent or scheduled in disruptive intervals, time-consuming tracking and reporting activities, or other day-to-day expectations that divert attention from their ability to focus on development. As engineering managers, we can support our teams by finding ways to reduce these types of process friction. Remove this friction step by step by identifying, understanding, and then remediating.

To remove friction, you first need to see it. Listen to your engineering team's perspective since they often have a point of view on whether the processes of a project are too time-consuming. Pay close attention to how they spend their day, how much time is devoted to development and other activities, and whether those other activities are generative or not. Generative activities might include pairing with a cross-functional partner to work on a solution to a project problem. Non-generative activities would include things such as daily standup meetings. Look at the ratio of development to non-development activities and see whether it makes sense to you intuitively. It may be understandable to spend more time in meetings at the beginning of a project, but as it goes along, you expect this to taper off as the focus shifts to delivery and execution.

If you identify areas where a process is infringing on productivity, the next step is to understand what purpose the process is serving and why it has become a burden. Determine whether this is a case where the process is good but has drifted in a problematic direction, such as a daily standup that is taking too long, or whether the process itself may not be the right approach, such as logging time into a system daily.

Remediating process friction may involve direct intervention or solutioning and negotiation. For example, if standup meetings are taking too long because a concise format is not being followed, you may intervene and correct that directly. You could say during the meeting *I want to make sure we are staying focused on the task at hand and not getting off track, so I'm going to start reinforcing the meeting format. State your focus for the day and any blockers you have only. Jane, focus and blockers?* On the other hand, if the situation involves other process stakeholders, partner with them on solutioning and negotiate an answer. Connect with them one on one to let them know your concerns about how the process is impacting productivity and that you would love to find a way to fulfill the process goals without slowing down delivery. Have some ideas in mind to talk through with them but listen to their point of view first.

Removing people friction

Another source of project friction can be the people that your team interacts with. This might be coworkers dropping by the desks of engineers, frequent emails seeking updates or other information, micromanaging stakeholders, or distractors from unrelated projects. Time to focus on development is important for maintaining momentum and minimizing broken concentration from context switching. Once again, engineering managers can identify, understand, and remediate this friction.

Friction from people can sometimes be hard to see, especially in remote working environments. This friction can go undetected when engineers attempt to be helpful to a range of different people and keep everyone happy, only to end up losing some of the momentum in their own work. Identify this friction by asking probing questions during your one-on-one meetings with engineers:

- Who from the project team do you hear from directly?

- Who are you fielding questions from day to day?

- How much of your time do you think is spent on other projects? How is it spent?

These can help open the dialogue, and you can probe further from there to get a sense of whether there is friction or not.

Once identified, this type of friction is often easy to understand. Maybe someone is dropping by for ad hoc updates because they are not in the right communication circles and you can simply add them. Maybe people are seeking information on another project because it is not well documented, or they don't know where to find the documentation, and you can solve that. Maybe stakeholders are micromanaging because there is something lacking in the communications you are providing, or perhaps they feel like they don't have a forum to contribute their ideas and you can solve that.

One of the easiest ways to solve people friction is, somewhat ironically, adding more processes. Sweeping actions that address a problem with a specific person by informing them (and everyone) how to get needs met with a process can provide a clean and blameless resolution. For example, you might establish office hours for ad hoc questions and communication and let everyone know that outside of those hours, the team needs to focus on development. Or the opposite, you could introduce quiet hours as a period of time each day with no meetings and no interruptions allowed so that the team may remain deeply focused on their deliverables.

Removing tooling friction

Friction from tooling occurs when the tools your team relies on to develop and deliver their work are slow and tedious or when there is not enough automation in the development life cycle. This may be lengthy builds, delays in getting updated code into a test environment, manual checks and gates such as bottlenecks in code reviews, or any other time sinks from development tools or the lack thereof. This type of friction is usually completely in the hands of the engineering team to resolve but may involve considerable effort to do so. Once again, engineering managers can look to identify, understand, and remediate this friction.

Tooling friction may be painfully obvious in your project, particularly if you are contributing to the development. Engineers are often vocal about their struggles with this friction, but you can still ask direct questions of the project group about their biggest day-to-day bottlenecks. Sometimes this friction is not as noticeable as a time sink but is an opportunity to move much faster, such as revising a frequent operation to run in milliseconds instead of a minute.

Once you've identified an opportunity to help the team move faster with tooling, it is usually easy to understand why it is occurring, but it can be helpful to quantify the impact. Since resolutions to tooling problems can be time-consuming themselves, it is useful to have a sense of the current time cost of a tooling problem before you think about possible solutions. Use simple back-of-the-envelope calculations to determine how much time this friction is costing the development team on average. For example, if builds take 30 seconds and are run on average every 20 minutes by 6 developers on your team, you can calculate that you are losing roughly 54 minutes per day on builds if your team members each code 6 hours per day.

Remediation of toolchain friction may involve adding or switching a CI/CD component or moving more of your conventions into code or tests. In general, the efficiency of automation has a big impact on project delivery, so if anything can be automated, it is beneficial to do so. Moving more of your conventions and policies into tests or pre-commit hooks not only speeds up those checks but also removes conversations and emotions from the delivery process.

Now that we have learned the secrets to smooth project delivery, let's go over some specific problems that may crop up during your project.

Project problems and solutions

Every project has its share of speedbumps or unexpected events. Here we will introduce two of the most common project problems along with different methods of resolving them.

You need to do more with less

Sometimes you need to find a way for a project to do more with less. For example, you might reach the end of estimations, and when you present them to your leadership team, they say *Great, you think you can do it in 6 months—we need you to do it in 3.*

One approach to this is to use the project management triangle. This method involves balancing three constraints: scope, time, and budget. Sometimes this is stated as *You can have it good, you can have it fast, you can have it cheap; pick two.* So, for example, if you need it fast, you would aim to negotiate a reduced scope or an increased budget.

In practice, you can often come up with additional levers and compromises to do more with less. When reducing scope is not an option, consider time, staffing, phasing, and technology choices.

Time

More time may or may not be an option on your project, but you can obviously achieve more with more time allotted to work on something. This is a useful lever if you are working with limited staffing.

Staff

If you need to shorten a timeline, adding staff can help up to a certain point. You may be able to augment your team with engineers borrowed from another team or contracted from an agency temporarily. This lever only works to a limited extent; you can't keep adding engineers indefinitely since there is a cost to bringing them up to speed and coordinating efforts. Increasing staff works best when you can add a small number of engineers who already have some familiarity with the product.

Phasing

Another option to do more with less is to negotiate phasing. This is where it still takes the full estimate of time to deliver the complete project, but some features and functionality are released earlier in phases. This provides a compromise of some features in a shorter timeline, along with enough time to complete the work.

Technologies

Lastly, you may look at simplifying the implementation to make a project more achievable. It may be possible to deliver a solution more quickly by using off-the-shelf software components that you can install and configure to work as needed. Depending on the project and product, this can be a very effective means of doing more with less, where the main trade-off is less flexibility in the future.

Now that we have learned how to use levers to solve difficult project scenarios, let's move on to when the project scope is a moving target.

You have scope creep

At the beginning of a project, you agree with stakeholders on a set of requirements and a delivery plan. Sometimes, during project delivery, the set of requirements for the project continues to grow without any change in the delivery plan. This is known as **scope creep** and it is a common project problem.

Changing scope as product ideas evolve is not necessarily a bad thing, but when timeline and estimate expectations are not permitted to evolve along with them, then you have a problem. Manage this by using your estimate assumption, negotiating trade-offs, and raising risks.

When we learned about estimating earlier in this chapter, we introduced estimate assumptions. They may seem excessive initially, but one of the main purposes of estimate assumptions is to anticipate and preclude scope creep. Assumptions will not stop scope creep from happening, but they will provide a concrete written record of what is out of scope that can form the basis of a discussion. Most stakeholders will be understanding of this when presented with the planning document that they have agreed upon and then be open to negotiating different resolutions.

You can also negotiate trade-offs on the fly as scope creep is encountered. This can be powerful since it gives stakeholders the option to have a change they want with no resistance and gets them into the habit of thinking about features as trade-offs. For example, you may be approached with an idea:

Stakeholder: *We would really like to include a page that lists all of a user's posts in the upcoming release—is that possible?*

Engineering manager: *Absolutely, we can trade something out of the release. What would you like to deprioritize so that we can fit that work in?*

If you are still beleaguered with scope creep after trying these techniques, you can capture it in your list of project risks and publicize it to all stakeholders and leadership. Scope creep is a project risk, and if you haven't been able to resolve it, you must start being vocal about it. This allows the wider leadership to consider the impacts of the scope creep and prepares for future difficult conversations you may need to have, such as *Why are we behind schedule on milestones?*

Summary

In this chapter, we learned how engineering managers can lead their teams through project planning and delivery by doing the following:

- Begin setting the stage for the project by understanding your environment and what you must provide to accommodate and support your team in that environment

- Finish setting the stage by evaluating and capturing opportunities to advance your engineering goals during the project

- Produce effort estimations with awareness of the amount of flexibility permitted in your plan, then break down work into small units and capture any assumptions you are making that impact the effort

- Prioritize the features of your project with stakeholders so you understand their relative importance and know what may be dropped if needed

- Address potential risks in the project by identifying them, organizing them into a matrix, prioritizing them, communicating them, and remediating them

- Put together a roadmap or timeline that captures sequential needs and dependencies

- Formalize lists of stakeholders, including those on the fringe, contributors with their roles and ownership, and a communication strategy

- Start project delivery with a kick-off that sets a project vision to align the team and give them a sense of purpose in their work

- Ensure your user stories or tickets have an approachable size, thorough acceptance criteria, and the right framing information

- Look for and remedy sources of friction or obstacles in the project caused by processes, people, and tooling

- Use time, staffing, phasing, and technology choices as levers to find solutions to accomplish more with less

- Address scope creep with your project estimate assumption by negotiating trade-offs and by raising risks

With that, you are ready to deliver and release your project to production. In our next chapter, *Supporting Production Systems*, we will learn how to provide for the day-to-day needs of a newly launched product or feature.

Further reading

Project methodologies and estimation

- *Project management in software development* (https//hackernoon.com/project-management-in-software-development-key-questions-answered)
- *Project management methodologies cheatsheet* (https//afroant.com/2021/08/18/project-management-methodologies-a-cheat-sheet/)
- Estimation checklist (https//github.com/togakangaroo/estimation-checklist)

Prioritization

- In *Chapter 14*, see the section on prioritization for an introduction to different methods of prioritization and how to choose a method
- *How to prioritize using these 9 mental models* (https://www.ejorgenson.com/blog/how-to-prioritize-using-these-9-mental-models)

Risk assessment

- See *Chapter 11* for a further introduction to managing risk
- *How to run a pre-mortem* (https://www.antmurphy.me/blog/2023/4/13/pre-mortems)
- *Klein, Gary. "Performing a project premortem."* IEEE Engineering Management Review 36.2 (2008) (103-104. https://www.researchgate.net/publication/3229642_Performing_a_Project_Premortem)

Roadmapping

To cover a range of different scenarios, here are three different approaches to roadmapping:

- *Evidence-based scheduling* (https://www.joelonsoftware.com/2007/10/26/evidence-based-scheduling/)
- *How to build a product roadmap everyone understands* (https://www.prodpad.com/blog/how-to-build-a-product-roadmap-everyone-understands/)
- *On writing product roadmaps* (https://goberoi.com/on-writing-product-roadmaps-a4d72f96326c)

Requirements

- Functional versus non-functional requirements (`https://www.geeksforgeeks.org/functional-vs-non-functional-requirements/`)

- *Patton, Jeff, and Peter Economy. User story mapping: discover the whole story, build the right product.* O'Reilly Media, Inc., 2014. (`https://www.oreilly.com/library/view/user-story-mapping/9781491904893/`)

General project delivery

- *Brooks Jr, Frederick P. The Mythical Man-Month.* 1995. (`https://www.informit.com/store/mythical-man-month-essays-on-software-engineering-anniversary-9780201835953`)

- *Cagan, Marty. INSPIRED: How to create tech products customers love.* John Wiley & Sons, 2017. (`https://www.wiley.com/en-us/INSPIRED%3A+How+to+Create+Tech+Products+Customers+Love%2C+2nd+Edition-p-9781119387503`)

Supporting Production Systems

In software development and operations, **production systems** are the instances of applications and services that are actively in use by end users. Once we build and release a network-accessible software system, it is said to be running in a production environment rather than a development, staging, or test environment. The activities that engineering and operations teams engage in to keep production systems available and troubleshoot operational issues are collectively known as production support.

Production support is an area with vastly different practices and norms depending on the company, industry, and customer expectations. In large technology organizations, it may be highly formalized across the company as **site reliability engineering**. In **software-as-a-service** (**SaaS**) settings with business customers, there are typically contracts with specific performance metrics that must be adhered to. Consumer product companies may vary widely in their approach, with some putting more thought into planning the initial build of software than into production support. Since the performance of your production systems determines your end user experience, businesses that operate these systems must provide sufficient service to retain users and survive. Engineering managers have a central role in fulfilling this need.

When we think about supporting production systems, we might immediately start thinking of specific techniques (e.g., observability and monitoring), targets (e.g., service-level objectives), or procedures (e.g., on-call response). Each of these is a useful tool in your support plan, but an engineering manager will produce the best outcomes by first giving the work meaning.

In this chapter, you will learn how to give meaning to the work of supporting production systems by building a culture of individual commitment to reliability. You will learn why this matters and what happens in production support when you don't have this culture. You will learn how to activate this culture and how to mobilize it by raising awareness with performance and robustness signals. You will learn how to introduce techniques, targets, and processes that make production support run as smoothly as possible. By the end of this chapter, you will be equipped with an understanding of how to create a self-improving reliability-oriented engineering team.

This chapter is divided into the following main sections:

- Creating a commitment to reliability
- Raising awareness of reliability
- Reliability solutions

Let's get started by learning how and why we give meaning to supporting production systems.

Creating a commitment to reliability

Supporting production systems can be some of the most stressful work we do as software engineers. It may involve late nights, spoiled weekends, or interrupted family occasions when engineers must dig into incidents for hours on end with half-asleep brains, all while knowing the company is losing money by the second until systems are back online. It is crucial work that can be incredibly unpleasant and disruptive by its very nature. We may strive to make incident and support scenarios easier to manage and resolve, but in most settings, we can never shield our teams from them completely. In most engineering teams, production support is inevitable.

The goal of production support is to find a balance between the inherent stress of the work and the reality that systems must remain online and available. Our objective is to support these systems in such a way that we avoid burnout while delivering the best possible level of service.

Because this work is challenging is exactly why it gains the most benefit from having meaning. You have the authority to establish mandates and directions for supporting production systems. You can tell your team what is required, what they must do, and when they must do it. But your engineers decide for themselves whether or not they will follow that direction, with what level of commitment, and how they will feel about it. Does this work feel like a burden they are saddled with from above or a challenge they conquer from within? This will depend upon the framing you give it. Great engineering managers give work meaning.

There is a popular abbreviated quote on motivation from Antoine de Saint-Exupéry:

"If you wish to build a ship, do not divide the men into teams and send them to the forest to cut wood. Instead, teach them to long for the vast and endless sea."

The full translation is closer to the following but has the same sentiment:

"One will weave the canvas; another will fell a tree by the light of his ax. Yet another will forge nails, and there will be others who observe the stars to learn how to navigate. And yet all will be as one. Building a boat isn't about weaving canvas, forging nails, or reading the sky. It's about giving a shared taste for the sea, by the light of which you will see nothing contradictory but rather a community of love."

—*Citadelle*, 1948 (`https://quoteinvestigator.com/2015/08/25/sea/`)

In that sense, we can say production support isn't about monitoring or automation or incident procedure; it's about instilling a commitment to reliability and a desire to strive for it. While the sentiment of this quote could be applied to many scenarios involving alignment and motivation, it is particularly applicable to the challenges of production support, which benefits greatly from individual creativity, self-organization, and personal determination.

When engineers are not personally committed to reliability, they will follow the letter of given expectations while not believing in the spirit of them. They will be more likely to sleep through an alert or avoid inconvenient duties. They will be less likely to spend time hardening systems and working toward long-term reliability improvements. They will be more impacted by the stress and frustration of production support and take a more negative view of it, potentially spreading that view to those around them and eroding team morale. When engineers lack personal commitment, teams are doubly impaired by reduced effort and reduced morale.

For more on the psychological basis of motivation, jump ahead to *Chapter 9*.

Now that we understand what we need and why we need it, we can examine how to build this level of commitment to reliability in our teams. We will do this by creating a culture of ownership and pride in the work and by communicating to engineers how their work is making a difference.

Ownership mindset

As engineering managers, we want to build commitment first by creating a sense of ownership for our systems. In this context, an ownership mindset is the feeling that something belongs to you and that you are responsible for it. Your workplace and team may or may not have a clear structure of ownership for its technology systems, but either way, it is worth the effort to establish it as much as you can in your team.

When you don't feel like you own something, you don't feel responsible for it; in other words, you don't feel a personal commitment to it. You may be asked to look after it, but you still believe in your mind, "*This is not mine; I don't really get it. I'm just going to do what I have to do and then hand it off soon.*" In this scenario, most people will do the minimum to meet expectations or improve things marginally but are not likely to take a long-term view of the system's health, needs, and best interests.

Developing an ownership mindset about something generally follows from becoming personally invested in that thing and its success. Becoming invested is the product of time, involvement, and having a share in decision-making. We become invested in the things we build ourselves or steward closely over time. We care about them more and personally desire them to reflect the work we have put into them. This is one of the reasons that dedicated product engineering teams work well because they have the time and opportunity to become invested in a particular area. It is hard to become invested in anything when you move from product to product or system to system in a short period of time. You simply don't form the same level of attachment. Similarly, we need our own ideas and choices to be incorporated into the work to feel invested.

In addition to investment, an ownership mindset needs finite bounds. Ownership works best with defined edges. More senior engineers can generally handle larger areas of ownership, but there are always limits. Team environments where everyone owns everything may devolve into the feeling that no one owns anything since it is simply too much to keep track of and understand. Ownership often follows product lines, but as the engineering manager, you will need to determine what the right size, area, and divisions are for your team.

Lastly, an ownership mindset follows from the ability to understand the systems in question. People resist taking ownership of things they don't understand or don't feel comfortable operating. Engineers need access to the knowledge and resources required to feel confident in owning a system.

With the time to become invested and the opportunity to focus on an area and understand it, your engineers will develop an ownership mindset that fuels their commitment to reliability.

Taking pride in the work

The next step to building a commitment to reliability is instilling a sense of pride in the work being done. Pride can be good or bad in different contexts, but in the context of building commitment, it is essential. In this context, pride is the feeling that we have put something into the world that meets or exceeds expectations, and that thing is a positive reflection of our efforts and abilities. It is good to take pride in our work.

Since all engineers strive for mastery of their tools and techniques, the output of our work forms a reflection of our efforts and influences our self-image. Our commitment to the work reflects our commitment to ourselves and creates a positive feedback loop where we grow our confidence as engineers by delivering work we are proud of. Confidence is our reward for getting through the challenges of our work. If we are not able to take pride in our work, we miss out on that positive signal, lose the reward of feeling like we are growing, and lose enthusiasm and commitment to our work.

The fuel we need for the positive cycle of growth and pride is reputation and recognition. Given we have the foundation of an environment where we are challenged and respected by our peers, developing pride in our work follows from earning peer respect and receiving appropriate recognition. We want confirmation from those around us that our work is good through conversations and direct feedback such as pull request reviews. Then we want some form of recognition of our accomplishments. This recognition may be given in different ways, such as verbal acknowledgments, incentives, public accolades, or promotions.

When we are in a mindset of taking pride in our work, it feels like a personal affront to us when that work fails or underperforms. We feel the personal desire to deliver reliable systems because we see ourselves as the sort of engineers that do that. We don't think about reliability because of a rule; we think of it because failure hurts our pride as engineers.

With encouragement and reinforcement to take pride in the work, engineers will develop a personal commitment to reliability.

Making a difference

The last foundational piece of building a culture that cares about reliability is reinforcing the sense of community, that engineers are helping each other and their efforts are making a difference. A sense of community is what happens when we feel like we are a member of a group and we want to live up to the expectations of that group. We share in the group's hardships and successes. We feel empathy for the group members, and we want to help them. We feel like we are needed, and our potential failures would be hard on the group, and we don't want that for them.

If we feel like our work is not making a difference, it is hard to stay motivated through difficult or unpleasant tasks. We think, "*Why am I suffering through this when it's not going to make any difference to anyone?*" Even if we know it is something we are supposed to do, we probably won't approach it with the same determination that we would if we had greater confidence in the tangible impact of the work.

Building a sense of community in your engineering team takes time and doesn't usually happen overnight. It helps if you are in a scenario where you can work with the same team members and allow them to develop bonds with each other. You can encourage the bonding process by organizing team social activities, having lunch together (even over video conference), celebrating achievements together, encouraging humor, and gamifying team objectives to create lighthearted competition. Practices such as these that help your team get to know each other better will speed up the process of developing a sense of community in the team.

Reinforcing that work is making a difference requires consistent follow-up and communication with your team. It helps to verbally acknowledge support efforts and reliability improvements made by your team. Go a step further by finding out and communicating how your team's efforts have impacted the business and the lives of those outside your team. Bring these up in group settings and congratulate them on the impact their work is making.

Having empathy for their community and visibility into how their work is making a difference gives a sense of purpose to this challenging work. With an ownership mindset over their systems, pride in their work, and the belief that they are making a difference in the lives of those they care about, you have created a culture of personal commitment to reliability on your engineering team.

With this culture of commitment, you only need one more element to produce an engine of reliability improvement, and that is awareness of the performance of your production systems.

Raising awareness of reliability

Entire industries have spawned from the concept that awareness is an effective catalyst for action. A classic example is wearing a step tracker throughout your day to count how many steps you have taken. These wearable step-counting devices were created with the premise that if you see how many (or how few) steps you have taken in a day, you will be naturally motivated to increase or improve that number. You may layer gamification on top of awareness to further incentivize goals, but it seems that awareness alone is enough to inspire action when the desire is already there.

So, when our goal is supporting production systems and improving their reliability, it follows that we may create the conditions for transformative action just by raising awareness of the performance of systems. To this aim, anything at all we can do to raise awareness of the state and trends of our production systems will trigger a series of actions that will have a positive effect on their reliability.

With this understanding, our efforts as engineering managers can focus on how we can raise awareness of systems' performance and robustness.

Metrics

Information we collect on system performance is usually referred to as **metrics**. Metrics is a general term used to describe various raw measures of resource utilization and behavior, such as latency and availability. When these metrics are aggregated together in meaningful ways, we can use the information to gain insight into system performance, bottlenecks, and potential areas for improvement. They help engineering teams understand systems' health and the impact of changes they are making and identify potential problems before they become system failures.

Collecting metrics from different sources is known as monitoring; we will talk about monitoring in the *Reliability solutions* section later in this chapter.

Communicating metrics actively

Active communication of systems performance means directly and personally providing your team with verbal or written metrics updates. This information is useful, but because it is delivered personally, how it is received depends on the tone and framing of the updates. With a positive tone, these updates can be exciting and energizing. With a negative tone, they can be discouraging and demotivating. The goal of active communication is to deliver it in such a way that the team is curious and inspired, not attacked or humiliated. These are strong words, but as we learned in the last section, when engineers are committed and invested in the work they do, perceived failures can be incredibly personal and emotional. This is why it is important that, especially with direct communications, we frame them with care and empathy. Present problems as shared challenges; for example, if we were to communicate about the performance of a new product, we might say, "*While we're on the topic of performance, there is also our new API endpoint preparing for launch with a 200-millisecond average response time. This is a great optimization opportunity for us to see whether we can cut that in half prior to launch.*"

There are many potential forums for actively communicating performance metrics. You may feature them in incident post-mortems, technical deep-dive presentations, or email newsletters. You can communicate them internally to your engineering team, and you can communicate them more broadly across teams if that makes sense. In general, it's good to communicate metrics to your engineers more frequently and communicate them widely less frequently. This is so that you capture the general trends and highlights for the wider audience and not the day-to-day fluctuations.

Communicating metrics passively

On the other hand, passive communication of systems performance means placing information where it will be observed regularly by your engineering team during the course of their day without any further action other than the initial placement. This form of communication can be very powerful when the information is placed in such a way that it can be seen many times a day, every day. The goal of passive communication is to make the information understandable and conveniently placed.

Metrics are not actionable if they are not understood. Display metrics at a meaningful level of granularity and on a meaningful timescale so that engineers can see relative scale, anomalies, and recurring issues. Make sure your metrics are aggregated and delivered in such a way as to produce understanding and insight.

Finding the best location for passively communicating metrics can be an exercise in creativity. If your team spends time in an office together, setting up metrics dashboards on dedicated displays placed around the seating areas is a great option. This ensures metrics will be seen often and casually discussed, creating awareness and an interest in understanding them better throughout your team. If you are not in an office with your team, metrics dashboards are still useful; you can encourage your team to keep the dashboard on one of their monitors at home. Refer to the dashboard frequently to keep it top of mind and generate interest from members of your team. Another great option is to integrate metrics notifications into your group chat room. For example, you can have a chatbot post your performance metrics periodically.

Now that your team has a commitment to reliable systems and awareness of the state of systems, they will be looking for tools and solutions to improve performance and robustness. Your role is to provide them with the best resources you can to anticipate and identify issues.

Reliability solutions

Reliability solutions include software platforms, configurations, integrations, automation, practices, and procedures. They provide insights, debugging tools, and timely information to engineering teams. Depending on your organization, you may already have access to a wealth of resources to increase reliability, or you may need to chart your own path.

Numerous volumes could be written on approaches and options for instrumenting systems, so here we will give an overview of the concepts for engineering managers to be aware of. These include service objectives, documentation, monitoring, alerting, and service interruption procedures.

Service objectives

If your company operates in a business-to-business context or provides SaaS, you may have specific **service-level agreements** (**SLAs**) and **service-level objectives** (**SLOs**). SLAs are contracts with customers that outline the performance expectations of a system. SLOs are the specific target ranges of different performance measures that are included in the SLA. SLOs are often expressed as a range of less than or equal to a target amount. For example, *99.5% of requests will be completed in ≤250 ms*, or *99% of tickets will be resolved in ≤4 hours*. Establishing granular measures of acceptable performance can be useful even when formal contracts are not in place.

If your system serves customers or even if you just believe your team would benefit from a concrete goal to rally around, writing SLAs and SLOs can be a great way to capture those goals. They align and formalize expectations around the most important target measures of your team's production systems.

When writing SLOs, the aim is to capture only the relevant metrics in the usage and performance of your system. These metrics are sometimes called **service-level indicators**. To choose the right indicators, think about what you and your users care about and what affects and constitutes a good quality of service. Then outline an acceptable range for those measures as SLOs. The right targets need to find the balance between what your users want and what is realistic and consistently attainable.

Support documentation

Good support documentation is a foundational resource for ongoing production support work. It decreases the learning curve of the work, reduces the need for further escalations, reduces stress in the moment of response, reduces confusion and mistakes in performing work, and helps distracted brains recall vital information. Assembling complete support documentation includes the following as a minimum:

- Historical information and context
- Support goals or SLAs
- Troubleshooting steps
- Known failure modes
- Severity levels and their expectations
- Systems owners and contact information
- Escalation paths

Each of these serves an important purpose in helping engineers support production systems. You can add more resources to this list by asking yourself and your team, "*What would I like to know if I'm trying to fix this system at 3 in the morning?*"

Let's go through each of these documentation components and the purpose they serve.

Historical information and context

To know where you are, it helps to know where you are coming from. Including documentation that explains why a system or component works the way it does can be an immensely helpful contextualizing detail. It allows engineers to better understand the current state of the system and to reason better about incidents with that system. This information also helps engineers plan further evolutions to the system that may make sense. This information is also useful to preserve for future engineers that may one day work on the system.

Support goals or SLAs

If you have SLAs and SLOs, they should live within your overall support documentation so that anyone reading your support procedures can see them. If you have not formalized SLAs, capture a few broad target measures as indicators of performance problems and the need for further investigation. These will be needed to determine alerting thresholds for the system.

Troubleshooting steps

Regardless of how well understood the steps may be, capture the diagnostic procedures for all of your systems and include them in your documentation. These are useful as support checklists to help tired brains and new support responders alike.

Known failure modes

If your systems have any known issues or failure modes, capture them in your documentation along with the steps to resolve or restore. This is another useful checklist to support the range of engineers who may be trying to understand what is happening with a system.

Severity levels and their expectations

Severity levels specify the conditions that constitute a support incident, how urgent it is, and specific response expectations. Defining severity levels is an important guideline for what needs to happen in the event of a production incident. It helps safeguard against engineer burnout by letting engineers know when a response must be immediate and when it is safe to wait and respond to an issue at a more convenient time.

Systems owners and contact information

Sometimes when an engineer needs more information or cannot resolve an incident on their own, they may need to contact a system owner. Whether inside or outside of work hours, you don't want the barrier to a resolution to be ignorance of who the appropriate person is or how to contact that person. Owners and contact information can be documented as individuals or as groups with mailing lists. It is often useful to list both casual channels, such as a Slack room for questions, and urgent incident contact information.

Escalation paths

Formal escalation paths specify in what order incidents should be progressively escalated to which individuals or groups. They provide a clear chain of contacts for use during urgent incidents and the timeframes for those escalations to be triggered.

Together, these records provide for the foundational needs of any engineer tasked with supporting your production system. Keeping these records up to date over time is critical to your team's success. Establish processes to ensure your support documentation is updated as needed, such as including documentation in project plans and booking reminders for you to confirm policies quarterly.

Monitoring

Monitoring is the collection of useful metrics through instrumentation and signal aggregation in order to better understand a system's health. Monitoring is widely seen as core to supporting production systems, so there is a vast ecosystem of monitoring tools and platforms available for different needs and preferences. Depending on your work context, you may have many monitoring tools and configurations available to you already, but in case you do not, we will introduce some key concepts.

Monitoring can be roughly divided into five different operations on metrics data: collecting, storing, transforming, visualizing, and responding or acting upon data. Monitoring tools and platforms focus on one or more of these activities:

- Collecting data is typically done through logs (instrumenting services to log useful information) and cloud infrastructure reporting

- Storing data is often done in specialized data stores such as time-series databases

- Transforming data is often done with search indexing and query interfaces to produce useful metrics views

- Visualizing data uses charting tools to present and illustrate metrics

- Acting upon data usually involves setting up integrations that trigger automatic changes or notifications

Depending on your resources and preferences, you may want to utilize several monitoring tools you configure to work together or an all-in-one observability platform. Another trade-off is whether to use self-hosted open source tools, managed commercial products, or something in between.

Cost and ease of use are often the key considerations in deciding the right approach to monitoring and observability. If your goal is to minimize cost and complexity, you might choose an all-in-one self-hosted approach, such as *Netdata*. If you need simplicity but a more commercial-focused product, you might choose a managed platform with support and commercial integrations, such as *Datadog*. On the other hand, if you are not afraid of doing your own integrations and connections, you might prefer assembling your own monitoring approach using tools such as *Elasticsearch*, *Grafana*, and *Graphite*. There are almost countless resources for composing your monitoring strategy, so take some time to review what is out there and might best suit your needs.

Good instrumentation empowers your team with essential resources to confidently maintain and improve production systems. With a monitoring approach in place, you can then determine how and when to use automated alerting.

Alerting

Alerting tools react to monitoring signals by routing notifications to incident responders. They are an essential part of the overall production support tooling ecosystem since they make your team aware of performance issues immediately. They are necessary for calling attention to failures before this notice comes from customers or clients unable to access a service.

In alerting tools, you configure what resources to check, how to check them, what interval to check them at, and how soon to notify whether the resource cannot be reached. Typically, each alert will follow an escalation path of contacts until the alert is manually acknowledged by a responder. Many alerting tools also provide some reporting capability, summarizing incident metrics such as resource uptime, alerting events, and acknowledgment time.

Many of the most popular alerting tools are managed online SaaS platforms that provide interfaces for configuring alerting, but there are also self-hosted options. Some monitoring platforms provide built-in alerting capability. A key consideration for alerting is to account for the possibility that if your systems are down, you don't want your alerting tools to also be affected by the same outage. Ideally, you want to be made aware that systems are down even in a total failure scenario. This is one of the reasons why managed SaaS options are popular since they are isolated from the rest of your infrastructure and can be managed in such a way that they provide failover in the case of major internet events, such as an AWS outage.

When you configure alerts, you will need to specify who receives them. A common approach is to plan an on-call rotation for the first level of incident alerts and include more senior leadership responders in the escalation path. **On-call rotations** are schedules that describe which engineers are on call to respond to alerts during provided date ranges. If possible, it is a good practice to limit the time that each engineer is on call. Being responsible 25% of the time or less is a good aim for work/life balance.

Consider your monitoring strategy, budget, and failover risks to choose an alerting approach that works for your team and your company.

Service interruptions

Incidents where a loss of service occurs are known as **service interruptions**. These may be partial degradation of services or full outages where services become completely unavailable. Depending on your business, unplanned service interruptions may range from inconvenient to revenue-impacting or business-critical. Clear communication around a service interruption is an important aspect of the overall support of production systems.

Planned service interruptions

Planned service interruptions may be scheduled when required for activities that necessitate some downtime, such as upgrades or migrations. Unless a system doesn't require continuous availability, planned service interruptions are usually a last resort. If it is possible to complete an objective with zero downtime, that option is generally preferable. If a service interruption is necessary, it is conventional to notify stakeholders and end users before, during, and after the event, as follows:

- Identify all audiences that may be affected by the service interruption. These may be stakeholders, your leadership, engineering teams, cross-functional teams such as infrastructure or data, end users, customers, clients, or others.

- Group the audiences together into meaningful groups for different levels of communication depth and detail. For example, you might divide the notification audience into internal and external in order to be more formal with your external customers and clients.

- Notify your audiences well in advance, if possible. Send an announcement of the upcoming service interruption one to two weeks in advance or even longer if it is likely to greatly impact audiences.

- Send reminder notifications as the service window approaches, such as the day before and 12 hours before. Find a good balance of how many notifications to send to be informative but not excessive.

- Notify your audience a few minutes in advance as you begin your planned service interruption.

- After service is restored, notify your audience of the outcome of the service interruption and any further relevant information.

If planned service interruptions are a regular occurrence in your work, adapt these steps to create your own written guidelines captured in your support documentation.

Unplanned service interruptions

Unplanned service interruptions require urgent communication so that those who are impacted are informed that you are aware of the incident and actively working to address it. These incidents need a less structured and more progress-oriented approach to communication.

Since unplanned service interruptions are generally inevitable in distributed systems, it's a good idea to capture in your documentation when and what communications you expect from your team including the following:

- Specify roles for a service interruption. It is generally a good idea to have at least two people responding to a high-severity incident, with one working on the resolution while the other handles the communication. It is very difficult to do both at the same time, and attempting to can make both efforts suffer.

- Provide contact lists of stakeholder audiences for each product or service. For example, you may want detailed incident communication to go to a list of engineering stakeholders, while product-level updates go to product stakeholders.

- Specify communication intervals for different severity levels and audiences. For example, you might want updates sent to internal teams at least once an hour.

- Include a communication template or example. It is helpful to illustrate, prompt, or define what characterizes good incident communication.

- Specify the timeline for sending preliminary and final **incident reports**. Incident reports summarize the cause, detection, steps taken, and outcomes of the incident. They may include recommendations for the future. You might want a preliminary incomplete report within a day and a final report within a week.

A common way to simplify incident communication is to create a status page. **Status pages** provide a known consistent web interface where your customers or clients can view or subscribe to updates about active incidents. They are convenient since they avoid having to manually keep track of who to update and how to update them. Many observability platforms offer status page add-ons, or there are standalone managed options such as `status.io`.

Summary

In this chapter, you learned how to support production systems by creating a commitment to reliability on your engineering team, raising awareness of the performance of your systems, and utilizing specific conventions and tools.

First, we learned how instilling our engineering teams with a personal commitment to strive for reliability is a powerful technique to reduce the stress and burden of difficult support work:

- Give engineers an ownership mindset by getting them invested in the work, sharing decisions, and growing their understanding and confidence

- Help engineers develop a sense of pride in their work by growing their reputation and providing recognition of achievements

- Help engineers see how the work they do is making a difference in a community that they care about

Next, we learned how to activate that personal commitment by raising awareness of the performance of our production systems:

- Use active communication periodically to raise awareness of trends in systems performance and underscore the importance of that performance

- Use passive communication to raise the day-to-day interest, understanding, and insight into systems performance

Finally, we learned many of the core concepts and conventions used to manage the day-to-day needs and occurrences of supporting production systems:

- Define SLAs and SLOs to capture and formalize target measures for production systems

- Maintain documentation that outlines the historical context, support goals, troubleshooting steps, known failure modes, severity levels, system contacts, and escalation paths

- Instrument your production systems with a monitoring strategy that gives you and your team a deeper understanding of the performance of your systems

- Integrate an alerting mechanism so that you know as soon as trouble strikes

- Set expectations of what your team should do in the case of service interruptions

With that, our production systems are well looked after, and our engineers are dedicated to further improving them every day. Now that we have mastered our technical needs, we will zoom out to learn how we can build better products by working better with our partner teams in *Chapter 7*.

Further reading

- Awesome incident response (`https://github.com/meirwah/awesome-incident-response`)

- Awesome status pages (`https://github.com/ivbeg/awesome-status-pages`)

- *Bellotti, Marianne. Kill It with Fire.* No Starch Press, 2021. (`https://nostarch.com/kill-it-fire`)

Part 3: Managing

In this part, you will learn how engineering managers foster great working environments. These chapters cover relationships, interactions, and team performance. You will learn how engineering managers work cross-functionally, communicate effectively, improve operations, build cultures of accountability, and manage risk.

This part has the following chapters:

- *Chapter 7, Working Cross-Functionally*

- *Chapter 8, Communicating with Authority*

- *Chapter 9, Assessing and Improving Team Performance*

- *Chapter 10, Fostering Accountability*

- *Chapter 11, Managing Risk*

7
Working Cross-Functionally

Engineering work often involves working closely with people who are not engineers. **Cross-functional teams**, where contributors from several different departments or skillsets collaborate on a shared effort, have become the primary configuration for software development work. These teams are arranged to have the right expertise to complete meaningful and impactful work together.

Engineers may work on cross-functional teams with a wide range of members. Contributors might include those from other engineering disciplines or those from design, data, operations, analytics, product management, project management, or business backgrounds. Many others may be a part of your cross-functional teams, each with their own skillsets, priorities, and perspectives.

Companies take different approaches to cross-functional teams. They may set up a functionally aligned reporting structure with matrixed cross-functional teams or they might have cross-functional roles report directly to a single manager. These teams may be assembled for a single project or dedicated indefinitely to a product.

In each of these assemblies, engineers must overcome unique situational obstacles in working across disciplines. The engineering manager's role is to demonstrate, facilitate, and support their engineers in overcoming these obstacles.

In this chapter, you will learn how to form productive and lasting cross-functional partnerships based on mutual trust. You will learn why this matters to you and your engineering team. By the end of this chapter, you will have a deep understanding of how to cultivate relationships and what to do when you are faced with difficult or adversarial cross-functional team partners.

This chapter will take you through the following steps:

- Demonstrating cross-functional leadership
- Understanding your partners
- Aligning with partners
- Building strong relationships
- Difficult partnerships

Let's get started by learning how and why we demonstrate cross-functional leadership to our engineering teams.

Demonstrating cross-functional leadership

For cross-functional teams, project and product success is defined at the team level, not at the functional level. In other words, engineers cannot be successful without their cross-functional teams. Where software development is a team activity, we must take a whole-team view of performance and success. As engineering managers, we may tell our teams this with our words, but the best way to impart the idea is by demonstrating it with our own behavior.

If we tell our engineering teams that cross-functional partnerships are important but we consistently place engineering priorities over those of our partners and don't make time for them, our engineers will see this and come to believe those partnerships are not actually that important. If we rely on lip service and good intentions rather than concrete actions and compromise, our engineering teams will do the same. Their relationships and product outcomes will suffer.

Leading great cross-functional teams starts with demonstrating great cross-functional leadership. This means treating our cross-functional partners as if they are our own team by understanding them, making ourselves available to them, aligning with them, building strong relationships, supporting them through challenges, and working with interdependence. The time we spend investing in these relationships is returned to us when we have fewer challenges and misunderstandings in our day-to-day work. Our engineering teams see the way we are working and follow our example, leading to better collaborations and improved product outcomes.

Now that we understand how to lead cross-functional work by example, let's go through each of these steps.

Understanding your partners

The first step to building strong cross-functional partnerships is to check your ego. We all know that engineers are awesome, but ask yourself, do you feel the same way about the partner teams you interact with every day? Do you know them? Do you genuinely respect and appreciate them for what they contribute to the team? Do you think of them as vital contributors or just people who are doing some tasks that you don't really value? Your attitude and beliefs about cross-functional partners can't help but leak into your words and actions to some degree. For this reason, the most successful collaborators will always be those that appreciate their partners for their unique experience, effort, and perspective. A welcoming attitude with a curiosity to deeply understand others is the best foundation for a good partnership. With this in mind, go about the work of getting to know your partners.

As you get up to speed on your cross-functional partners, go through an exercise of gathering the following information:

- What are all of your cross-functional partner teams?

- Who are your key contacts on those teams?

- What do they do in their roles?

- Who are their stakeholders?

- What do they care about?

- What tools do they use?

- What makes their jobs easier or harder?

For example, you may work with data analysts who write and run reports for marketing teams and care most about data integrity and consistency. The more you understand them, the more you will include them in your day-to-day thought processes.

You can gather this information by setting up one-on-one meetings or by sharing a coffee or lunch break. Sometimes it may feel a little awkward to reach out and start a dialog for this purpose, and if so, that is a moment to draw on your bravery as a manager. Growing as a leader and partner requires some vulnerability at times. Let them know you are interested in understanding their role better so that you can see the big picture. If you are new in your role or in working with them, mention that and that you'd like to have a productive partnership. Lean into your curiosity to ask questions until you feel you deeply understand what they do, why, and how.

Gathering this information has many benefits. It gives you insights and can reveal opportunities to make big improvements in workflows. It allows you to anticipate and plan for the needs of your partners. It makes you a better resource to your engineers in contextualizing needs and situations for them. It strengthens the cross-functional relationship and builds trust with partners by showing them that you are interested in them and value their perspective.

From this point, you can find a regular cadence to check in with your cross-functional partners to be aware of what's new with them and their teams. This can be a meeting, but it can also be as simple as periodically checking in with them, stopping by their desk, or sending a chat message to ask, "*Hey, how are you? What's going on in your world?*"

It might be the case that you have so many cross-functional partners that it seems impossible to get to know and remain in touch with all of them. If you are in that situation, prioritize your partnerships in order of closeness and impact so that you have a list to work through. Take the time you have available and schedule it on your calendar so that you can work your way through the list. Schedule follow-ups with your closest partners more often and work with fringe partners periodically. Schedule them as meetings or just reminders to yourself to reach out at that time.

Now that you have assessed who your partners are and initiated communication with them, your next step is to bring your cross-functional efforts into alignment.

Aligning with partners

Functional teams typically have a whole range of simultaneous efforts, intentions, and goals. Taking proactive measures to align with cross-functional partners is how we avoid conflicts and, further, how we start to achieve a more productive partnership. We align with our partners by adopting a same-team attitude, uniting team visions, providing clarity on roles, and providing an aligned structure.

Adopting a same-team attitude

The way to create an aligned cross-functional team is to start acting like a single team. Give the whole team a sense that they are in this together and they have each other's backs. That means that for everyone on the cross-functional team, their goals are your goals and their problems are your problems. Believe this and demonstrate it through your words and actions. Enthusiastically tackle the challenges of your cross-functional partners. Look for mutually beneficial solutions and approaches.

Make sure any competition within the team is healthy and done in good spirit. Healthy competition reflects good sportsmanship in team members' attitudes and actions toward each other. Avoid antagonistic competition, such as seeking functional credit for cross-functional successes. Work to share the credit for team successes and always celebrate as a cross-functional team. If your engineering team is singled out by company leadership, for example, be the first to say, "*Thanks, we worked hard on this but we absolutely could not have done it without our amazing partners from design, product, and analytics working side by side with us every day, so a huge thanks to them.*" Or, better yet, if the situation warrants it, "*Are you kidding me? These folks right here—they knocked it out of the park and set us up to do our best work too. 100% a team effort on this one.*" Be your partners' biggest supporters and cheerleaders publicly and vocally to promote a one-team atmosphere.

Uniting team visions

In *Chapter 5*, we learned how important it is to give work meaning and alignment by setting a project vision. Also consider that teams often work from more than one vision. Teams may have visions at different levels of granularity, such as product vision, departmental vision, and functional vision. For example, it might be part of the vision of the engineering team to evolve the architecture in a particular direction or move to a new structure of component ownership. Each functional partner you work with is likely to have their own set of visions from their function and team context. Functional visions may be a source of cross-functional disagreement if they are not understood and accounted for.

Take the time to ask about and understand the vision or broad aspirational goals of each of your functional partners. Share yours with them. Figure out whether they align with each other, and if not, is there something you can do to better align them? If you have an opportunity to incorporate cross-functional perspectives while creating your functional vision, that's even better. Invite your cross-functional partners to come to an engineering meeting and share their team's vision and goals with the engineering team. Do the same in one of their meetings, if possible. Consider holding a cross-functional offsite event or exercise. Taking these steps to develop a mutual understanding of cross-functional drivers is a great way to avoid surprises and bond teams further.

Now that you have a same-team mindset and aligned visions, you can further align the team by providing clarity on roles and decision ownership.

Providing clarity on roles

Teams move faster, experience less friction, and are better aligned with each other when they understand exactly what is and is not a part of their remit. As the engineering manager, you are in a position to ensure this is specified and agreed upon for the whole team. In *Chapter 5*, we included the role definition as an essential part of the project plan and introduced the RACI chart. This information is not only helpful in setting up projects for success; it also improves cross-functional relationships.

Every cross-functional team will experience some disagreements. These disagreements can become sources of division for a team if you haven't established clarity in roles and the level to which each role must be involved or consulted on a decision. This division can be avoided by setting expectations for the roles of each individual. Giving this clarity helps teams stay aligned, close, and friendly in spite of differences in opinion that are bound to occur.

There are many alternative formats to the RACI that you can also consider (see *Project contributor roles* in *Chapter 5*); the important leadership action is aligning expectations on what each person contributes to the cross-functional team so that members can focus their efforts on those areas, support each other, and avoid unnecessary conflict.

With that, the next step of alignment is to move from individual roles to clarity in peer relationships.

Providing an aligned structure

Another way to facilitate alignment on cross-functional teams is to provide structures to support cross-functional peer engagement, such as formalizing peer relationships and introducing cross-functional pairing sessions.

As a part of the role definition, you may introduce formal relationships between roles. You might plan for specific cross-functional peer owners for a portion of a project, for example, setting peer collaborators as named feature leads from product, engineering, and design. This sets the three of them up for a close partnership, working in lockstep through the solution and features.

Cross-functional pairing sessions are another way to encourage alignment and work through problems. With these pairing sessions, the focus may be on a problem, goal, or specific user story. The pair can work through approaches and will benefit from developing a greater understanding of each other's priorities and expectations. Remote pairing sessions can be organized with the use of screen-sharing tools.

With an aligned cross-functional team, we can shift our focus to deepening the relationships between team members.

Building strong relationships

Relationship building is core to working cross-functionally. Strong relationships in your cross-functional network help you get more done because you have trust, buy-in, and support from your partners. These things do not happen magically over time; they are the result of effort, attention, and investment in relationships.

Sometimes we may think relationships aren't as important as being good at our functional jobs and conducting ourselves fairly. After all, we're here to do work, not make friends, right? It is true that great work is the ultimate goal, but when we fail to invest in relationships, we miss opportunities to do our best work. This is because we are human and we cannot avoid our own social and psychological underpinnings. If we don't take the time to connect and understand each other better, we don't feel as comfortable or trusting, and we are less likely to take risks and share our true opinions. We make safer choices. We hold ourselves back instead of taking a chance and forging a new path. Doing our best work requires that we trust each other on a deep level.

Knowing that we need strong relationships to do our best work, it is the job of engineering managers to foster these relationships for themselves and their engineering teams. We took our first steps toward relationship building earlier in this chapter by understanding who our partners are and aligning with them. Now we can deepen those relationships by making ourselves available, further understanding our partners, deliberately building trust, and maintaining a feedback cycle with them.

Making yourself available

To strengthen your relationships, you must invest your time in them. You can't develop relationships without spending time on them in one way or another. It can be difficult to find time in a busy schedule, but it is worth the effort to do so.

The crux of making yourself available to your partners is to be there for them when they need you. Being there for others when they need you builds trust with them since they begin to see you as someone they can rely on. If you are struggling to find the time, you can timebox interactions on your calendar with a recurring meeting or by blocking out time for yourself to follow up with each partner. Be sure not to overextend yourself or provide a level of attention that is impossible for you to maintain, since some may end up relying on you too much or taking advantage of your goodwill.

Further understanding partners

Earlier in this chapter, in the *Understanding your partners* section, we learned how to build a strong foundation of cross-functional knowledge. As the relationship progresses, build on this foundation by continuously showing the same level of interest. As you partner with them day to day, attend meetings, and discuss ideas and plans, earnestly seek to understand their point of view and why they have it. Avoid jumping to conclusions or making snap judgments. Have a personal goal of deeply understanding their point of view both as a representative of their functional discipline and as an individual. Doing this consistently allows you to keep your finger on the pulse of what is happening in their department, understand their driving forces, understand their personality, and provide better context to your engineering team. It also deepens the relationship by consistently demonstrating that you are interested in them and what they have to say.

Helping partners understand you

Along with growing your understanding of your cross-functional partners, be willing to help your partners understand engineering better. Be a resource for your cross-functional partners to gain an understanding of engineering points of view, drivers, and processes. Where there is interest and opportunity, be their technical guide and help them to understand how products and features work. Many will appreciate the opportunity to expand their knowledge. If your partner misunderstands something, take the time to help them understand better. Develop your ability to frame concepts for different audiences so that those concepts can be understood and appreciated rather than coming across as boring, confusing, or extraneous. Doing this makes your partners see you and your team as helpful and builds trust since they better understand your motivations. It also sets a good example and expectations with your engineering team of how to support and educate across functions.

Seeking and providing feedback

It's a good idea to go beyond understanding and proactively exchange feedback with your cross-functional partners. Exchanging feedback respectfully and earnestly builds trust that you can resolve your disagreements together in a mutually beneficial way. In your one-on-one meeting or other time with them, talk openly about your goals and desire to have a great cross-functional partnership. For example, you might start a dialogue by saying, "*I want us to have a great partnership on this project, so I wanted to check in with you and ask for your feedback on how it is going working with the engineering team. Are there any challenges on your side? Is there anything we or I could be doing better? Or anything you miss from past collaborations?*" From there, you can explore what feedback they may have. Generally, they will reflect the conversation back to you at some point and ask what feedback you have for them. Aim to give feedback that is constructive, kind, and candid. If you have critical feedback, give it with care. For example, you might say, "*I am loving the visual designs we are getting and really enjoying bringing them together, but I am noticing that they are frequently coming through 2–3 days behind schedule, which is putting us in a time crunch and causing us to have to work after hours. Do you have a sense of the dynamic that is causing this for you?*" Be honest about your needs but be careful not to blame. Assume they have the best intentions and take a shared problem-solving angle.

If you disagree with feedback they have given you, don't just dismiss it; make every attempt to understand their point of view. Dig into why they have this feedback. Understand the underlying dynamics and drivers at play. If you can see it from their point of view, you can better determine how to explain to them why you disagree and what solutions you see that they may be missing.

With feedback that you agree with, follow through on addressing it. Work with them or with your engineering team on a solution. If it is a work in progress, keep them updated along the way. If it is something you can't resolve, don't ignore it and leave them hanging; tell them that you cannot fix it and why specifically. They will respect you more for your honesty.

Building strong relationships has numerous benefits for your team and company, but you may still find yourself in difficult situations from time to time, so let's learn some techniques for working through those situations.

Difficult partnerships

From time to time, all engineering managers and their teams will find themselves faced with a difficult cross-functional partnership. Despite our best efforts and careful practices, not everyone will respond to us in the way we hope they will. Since we usually do not have the luxury of picking our cross-functional partners, it is an important skill to be able to overcome these situations and find ways to continue to do our best work. When you are faced with a difficult cross-functional partnership, follow these steps.

Make your manager aware

It's a good practice to keep your manager broadly aware of everything you are working on and specifically of the challenges you are facing in your cross-functional teams. This is not to cast blame or ask for help, but instead to give them a warning that you see a dynamic happening that is not ideal and that you are working on it. This allows your manager to have insight into the efforts you are putting forth in your work while also informing them of your awareness and actions in case the situation is brought to them by others. You may be inclined to just handle it yourself, but the danger of this is if your manager hears about it from someone else first, they may be surprised in a negative way and wonder why you haven't mentioned it to them.

Talk to your manager in your regularly scheduled meetings to update them every step of the way. Inform them of the problems you see, what you think the underlying causes may be, and what specifically you are doing to improve the situation. You might say, "*The partnership with _____ hasn't been going as well as I would like; they seem distracted and are not engaging during meetings, then bring up issues we've already discussed after the fact. I think they may just be overcommitted. My plan is to start by talking to them about this in our next one-on-one meeting.*" Then your manager will be aware and also have an opportunity to provide additional context they may have or give suggestions.

Once you and your manager are on the same page, you can work on improving the situation directly.

Lean into the relationship

A good relationship is a close relationship. When a relationship is not going as well as you would like, first attempt to improve the relationship by understanding the person better. Work on turning your adversary into your ally. If they are distant and hard to read, use your relationship-building skills and earnest demeanor to win them over. Every person you meet is an opportunity to learn something new, so with that in mind, look for qualities in the person that you admire or are curious about and use those to develop a genuine interest in them. Use the following relationship-building actions:

- Work to get a dialogue going with them. For example, *"What do you think about this new feature they asked us to do?"*

- Relate to them to encourage them to open up to you. If you notice they get annoyed by something on the project, share your frustrations about that thing.

- Find shared goals to team up with them on. For example, *"You seem a little frustrated by _____. It has been pretty crazy. What do you think we should do?"*

You can also use the Ben Franklin effect, asking them to do you a favor so that they start to see you as a person they like (`https://en.wikipedia.org/wiki/Ben_Franklin_effect`).

On the other hand, rather than being distant, it may be the case that they are just underperforming, not keeping up with the team, or their work is suffering in some way. If that is the case, dig into what is causing this. Is it a skill gap or more of a time gap? Once you understand the underlying causes, you can work to address them by either using the feedback methods we learned in the *Building strong relationships* section or by helping them to address skill gaps.

Work more defensively

Next, to improve the situation with difficult partners, you can work more defensively. This means working in such a way as to protect yourself and your engineering team from the effects of challenging partnerships.

How you work defensively depends on the situation you are facing. It involves anticipating the types of problems you expect and introducing mechanisms that prevent, diffuse, or change the dynamics of those problems. Some useful techniques for this include leveraging the project plan, defining your engineering inputs and outputs, overcommunicating, and changing the project management methodology.

Leveraging the project plan means making use of the agreements in the plan to call attention to and push back on issues you are facing with your cross-functional partners. Key areas from *Chapter 5* include the estimate assumptions, risks, and the *Good user stories* section. When your project plan stipulates these three things, you have an excellent framework for addressing situations such as the late delivery of requirements, stories missing necessary information, or requests that are outside of the agreed-upon scope.

Defining your inputs and outputs is a means of establishing defensive policies and procedures for your team. With support from your manager, you can introduce rules or expectations for what your team will accept and what they will do. For example, if your cross-functional partners are asking for releases late on Friday afternoon or some other time you disagree with, you might introduce a policy of no releases on Fridays. Establishing policies for cut-off dates and times or lead times for different types of requests is a common practice to set expectations across functions. Clear policies remove ambiguity and help to set everyone's expectations in advance. This keeps things civil and provides a basis for when you need to say no.

Overcommunicating is another way of protecting your team in the case of a challenging partnership. In this context, it means being vocal with the entire team about what you need, expect, and are doing. When you are highly vocal, it helps shift the problem from a *you* problem to an *team* problem and gives you more support in the situation. For example, if your cross-functional partner is repeatedly late with their hand-offs to your team, you might overcommunicate when you need something by to stay on schedule, bringing it up in your planning meetings, standups, and written updates. "*I'm sure the design team is on top of this but I wanted to mention that we need the final designs by tomorrow to be able to deliver this feature on schedule.*" Be vocal in an earnest way, always assuming the best of your partners and not in an antagonistic or accusatory way.

Changing the project methodology is something you might do to work more defensively in the case where priorities are changing often or new requests are coming in frequently. If you are working in a style where work is planned and committed to such as Agile sprints, but your product partner is introducing new work mid-sprint, that is a situation where you can move the team to a different style of work that can accommodate those changes better. Kanban is a useful project management style for changing priorities, since engineers just continuously take the top stories off of the board rather than committing to them in advance.

Develop a sense of the challenges your team is facing in their cross-functional partnerships and figure out how to avoid or overcome them in mutually beneficial ways before they happen.

If all else fails, escalate

Finally, if a partnership continues to be a drain on the team's quality of work, you may consider escalating the issue to their manager or another senior leader. This is a last resort because it moves the relationship into a more adversarial territory and can come across as tattletaling, which may reflect badly on you. People have an inherent desire to have others bring their grievances directly to them, so going to their boss is an aggressive move.

You have to assume that the partner in question will likely find out that you escalated the situation above them. It is likely to put a strain on the relationship moving forward. There are absolutely situations where you should escalate, such as blatant negligence or HR violations, but you want to be sure you are prepared for whatever the consequences may be. Talk to your manager about your escalation plan before you carry it out.

In most difficult partnerships, you won't have to go too far into these steps to get them back on track, but it is good to be thorough in your project planning and communication just in case you end up needing to.

Summary

In this chapter, we learned how to forge productive cross-functional partnerships that enrich our work and help us grow as leaders. We learned how cross-functional relationships are essential to produce our best work since the end product is not only code but also the result of collaborative efforts. To do our best work in cross-functional teams, we must build relationships and establish trust with our peers:

- Lead by example in demonstrating to your engineering team how to engage with cross-functional partners.

- Assess cross-functional partners by gathering information to understand them and their needs.

- Invest time in cross-functional partners to improve cross-discipline workflows, anticipate cross-functional needs better, and build trust.

- Set up regular time in your calendar to make sure you are keeping up with your cross-functional partners.

- Align across functions by adopting a same-team attitude, uniting your functional visions, establishing clarity in roles, and creating peer structures at each level.

- Deepen relationships across functions in order to develop the necessary trust and mindset for creativity and innovation.

- Deepen relationships by making yourself available to cross-functional partners, understanding them, helping them to understand you, and exchanging feedback.

- Improve difficult cross-functional partnerships by making your manager aware, leaning into relationships, and working more defensively.

Now that we understand how to build strong cross-functional partnerships, we are ready to take a deep dive into communication and what makes it effective in *Chapter 8*.

Further reading

- *Cagan, Marty. INSPIRED: How to Create Tech Products Customers Love.* John Wiley & Sons, 2017. (`https://www.wiley.com/en-us/INSPIRED%3A+How+to+Create+Tech+Products+Customers+Love%2C+2nd+Edition-p-9781119387503`)

- *How To Lead When You Have No Authority* (`https://medium.com/swlh/how-to-lead-when-you-have-no-authority-9f22206356d4`)

- *Forming Stronger Bonds with People at Work* (`https://hbr.org/2017/10/forming-stronger-bonds-with-people-at-work`)

8
Communicating with Authority

For engineering managers, communication is a topic that is relevant to nearly every aspect of our work. From talking with our teams and working with stakeholders to how we name software packages and variables, the need for clear and thoughtful communication is present in everything we do. A firm understanding of best practices in communication is important for every engineering manager.

Communication is one of the few force multipliers in leadership. The same series of events, when communicated well or poorly, will produce vastly different outcomes for individuals and teams. Language has the power to motivate, encourage, convince, contextualize, or simplify. It can prevent, avoid, or resolve a difficult problem. Or if it is not used well, it may cause misunderstandings, division, frustration, and anger. There is no exaggeration when stating that poor communication can lead teams to ruin and lose them the support of their stakeholders and leadership. Because of this vast impact, communicating well is a key foundational skill for engineering managers to master.

In this chapter, we will introduce the fundamentals of good communication for engineering managers. You will learn key principles and practices for communicating effectively. Then, you will learn how to plan and structure your communications. You will also learn strategies for communicating with your engineering team and **managing up** to leadership. By the end of this chapter, you will understand how to leverage your communications to set teams up for success.

This chapter contains the following sections:

- Principles of communication
- How to structure your communication
- Communicating with your engineering team
- Communicating with your leadership team

Let's begin by introducing the general rules of thumb for communication as engineering managers.

Principles of communication

There are numerous writings and studies on how to communicate effectively. Our focus is on what aspects of communication are important for engineering managers to understand and master. To understand what we need to know about communication to produce the best outcomes, let's examine how and why we set expectations, assume the best, say no with care, have an audience perspective, maintain authenticity, and give feedback with radical candor.

Setting expectations

The underlying cause of most frustration and conflict can be traced back to an expectation mismatch. In other words, when what we believe turns out to be false or is not upheld by those around us, it is almost always an upsetting moment. For example, if you believe elevator riders should be given room to exit before new riders get on and other riders don't respect that expectation, you are likely to be annoyed by the exchange because your expectations are not being upheld. In these moments, we may have emotional reactions. This dynamic is why setting expectations is core to good communication.

We have many beliefs—for example, something big, such as believing we are ready to be promoted to a new position, or it may be something small, such as which seat we occupy in the office setting, believing that seat is ours. Our beliefs often become deeply rooted expectations. As engineering managers, it is a key skill to recognize this dynamic and continuously manage those expectations so that they're in line with reality.

You may think that if you have omitted information, then people will reach out to you if they have questions or concerns. This assumption will not prevent troublesome expectations from arising. Don't make assumptions that people already share your understanding of a situation or aren't concerned with it. In the absence of deliberate expectation setting, people fill in the blanks on their own, guessing about the certainty, fungibility, or boundaries of a situation. Often, they will fill in the blanks with what they want to be true. If there is no communication to set expectations correctly, individuals may hold incredibly skewed and mistaken perspectives. Get ahead of this by clearly setting expectations in your communication.

Managing expectations is often an exercise in proactively communicating what you already know to avoid potential misunderstandings. Get in the habit of including the following concepts in your communication:

- **Certainty or permanence**: Communicate the likelihood of change and the potential magnitude of that change
- **Contingency**: Communicate any factors something may depend upon or that may lead to change
- **Purpose or criteria**: Communicate why something is happening and what the deciding factors are
- **Boundaries**: Communicate what is and is not included in the discussion at hand
- **Duration, quantity, or volume**: Communicate limitations of size or time

So, revisiting our earlier examples, we might set expectations about promotions for the engineering team as follows:

- **Certainty**: *This document outlines our current promotion process. This is a work in progress that we expect to revise continuously.*

- **Contingency**: *Promotions are contingent on budget availability and position availability. They are carried out twice a year after semiannual reviews.*

- **Criteria**: *Promotions are considered based on the following set of factors: _____.*

- **Boundaries**: *This information describes promotions for the _____ department.*

- **Quantity**: *There are unlimited positions available for senior engineers but only one position on staff currently for the principal engineer.*

Providing this level of detail up front sets key expectations so that engineers understand not only the performance expectations but also the broader context and organizational factors.

Even in informal communication, you can set expectations with this level of clarity. For example, *"Hey all, it looks like seating for our team is being moved to the southwest corner of the floor. You will receive an email this week with a specific seat assignment. Think of this as temporary for now. We should have longer-term seats determined by the end of next month. The goal of the new arrangement will be to reduce the distance between us and product managers and to minimize distractions."* With this, we have communicated the purpose, certainty, boundaries, and timeline.

Deliberate expectation setting is a powerful means of preparing your team for what may come. Now that you know how to prepare the team, let's see how you can encourage harmony by taking a positive perspective in your communication.

Assuming the best

In all communication, it is a good idea to start from a perspective of how you want the world to be rather than how you may suspect it is. In other words, give the benefit of the doubt and always start by assuming the best of everyone. This means assuming that everyone is honest, is doing their best, has good intentions, is good at their job, and has followed proper procedures. You can confirm this is the case in the course of your actions, but start from this perspective to avoid offending others, promote trust, and encourage the right behavior.

It can be incredibly damaging to offend someone by making a wrong assumption. At some point, everyone has been wrongly accused or singled out for something that wasn't their fault. When you jump to a wrong conclusion or blame someone for something they did not cause, it leaves a lasting bad impression on everyone involved. They will be less trusting of you in the future and more defensive. They may believe that you don't think highly of them and don't trust them. They may underperform because they think that they will be blamed, even if they have done everything right. Avoid this loss of trust by making a positive assumption instead of a negative one.

Good communication should promote trust. Assuming the best communicates to others that you trust them, believe that they have taken the right actions, and believe that they are working to the best of their ability. It encourages them to reciprocate trust by assuming the best of your actions and intentions as a manager. It demonstrates how they should treat others in a trusting and supportive environment.

Assuming the best also encourages the right behavior. Our expectations and beliefs about each other can become self-fulfilling prophecies. When we expect the wrong behavior and act as though we anticipate it, that behavior is more likely to materialize. When we behave as though we expect the right behavior, that behavior becomes more likely. Second-guessing people, micromanaging them, and treating them as though they are doing something wrong is going to demotivate and discourage the right actions.

You don't need to trust everyone blindly to assume the best; you can still confirm work in a supportive manner. For example, consider if someone hasn't shown up to work and it's a few hours into the day. Assuming the best would be to reach out to them with care and concern—*"Hey, are you okay?"*—rather than communicating something more accusatory, such as *"Hey, where are you?"* Assuming your coworkers have a good reason for anything they do is the right starting point. Another example is how you approach confirming a procedure has been followed. A positive assumption, such as *"You ran this by John, right?"* is better than expressing the same question neutrally; for example, *"Did you run this by John?"* Get in the habit of phrasing your communication in a way that expresses trust and positivity.

It is helpful to make positive assumptions, but sometimes, we need to deliver a negative response, so let's look at how to approach that.

Saying no with care

When you need to communicate no, it's a good idea to do so with care. No is a denial, a rejection, a statement that you will not give someone what they have come to you asking for. How this is handled can either cause frustration and strain relationships or produce acceptance and understanding. For the best outcome, take a step-by-step approach to saying no:

- **Make sure you understand fully before saying no**: What is this and why has it come to you? What are the underlying goals or drivers?

- **Exhaust every possibility for arriving at yes**: Think like an advocate for this person or project—is there any way it would be possible?

- **Make every attempt to help**: Is there a compromise that might satisfy the goal while being more achievable?

- **Assess the greater good**: What is the best outcome for the largest number of people? What is most fair?

- **Take a same-team perspective to frame the no as a shared problem**: Speak from the perspective of advocacy for this person or project.

- **Be clear on what is needed to get to yes**: Be specific in communicating what needs to change for future viability.

Following these steps helps avoid an unnecessary no, keeps you positioned as an advocate, and frames the situation as one you are in together as allies and not adversaries.

Let's say, for example, someone wants a piece of work done that your team doesn't currently have the bandwidth for. Why do they want this done? What purpose does it serve? What factors are involved? Are there other possible solutions to address the underlying need? Is there anything else you could do to help? What would be the impact if you did the work? What are the trade-offs between doing it and not doing it? If you conclude that it would be too harmful and disruptive to take the work on at this time, talk about why that is. "*I think this work makes sense and I would love to partner with you on this, but I have my team fully committed for the next two sprints. How about we email the product manager together and see whether they are willing to drop something so we can take this on? Hopefully, we can find a compromise, but if not, let's put it in the roadmap.*" On the other hand, let's consider that you think it's an unnecessary idea. In that case, you might dive deeper into the motivations and alternatives. What has led to this idea? Have they considered other approaches? What are competitors doing? What data has led them to believe that this work is important? Has this been prioritized by members of product and leadership? From there, you can be honest. "*It's an interesting idea, but it's hard to prioritize against the rest of the roadmap, which has a clear value proposition and empirical data supporting it. I think that to seriously consider this, we need to have a clearer substantiation of why it will be successful and justifies deprioritizing something else.*"

Since it is always hard to hear a no, deliver them with structure and empathy to help the message be understood. Next, let's discuss why it is important to think about the audience of our communications.

Having an audience perspective

There is a difference between telling something to others and having that information heard and understood. Many communications are lost on their audiences. We may think we have communicated well when in reality, our message was not received. This is why communication should be given from an audience perspective.

An audience perspective means recognizing who we are communicating with and tailoring our communication to be received by them. It means having an awareness that audiences are different and may have very different needs from communications. It means recognizing what their perspective is, what level of understanding they have, what they are interested in, and what is relevant to them. It means having empathy for when they are receiving this message, what else they might be doing, and how much time they may or may not have.

When communications fail to take an audience's perspective, we tend to ignore them. We may view them as tone-deaf, irrelevant, or uninteresting. We often tune out and don't receive them. They become buried in our inboxes. We look at our phones instead of listening. As communicators, this hurts us because it is in our best interest to have our work be understood, appreciated, and supported. Since we are all presented with information overload every day, the best communications need to cut through the noise.

From an audience perspective, our goal is to give people information that is relevant to them in a way that they can digest. That starts with understanding our audience and having empathy for them. Rather than sending one message to all audiences, we determine the right approach for each audience and frame the communication for them. This involves a little more work in drafting communications, but the investment gives us greater reach in our communications and results in less time spent on ad hoc updates, clarifications, confusion, and misunderstandings. Get feedback from your audience members to fine-tune your communications or find a trusted colleague to run it by before sharing it more widely.

In addition to considering the audience's perspective, let's examine how we can keep our communication rooted in the authentic realities of ourselves and the world around us.

Maintaining authenticity

Authenticity has become increasingly important in modern communication. It concerns how much people believe our communications are sincere. Our communications may be perceived as genuine or as merely saying what people want to hear. Regardless of our intentions, how our communications are perceived will greatly impact how people respond to them. To be seen as authentic, it is important to act with integrity, limit euphemisms, show your personality, and be vulnerable at times.

It should go without saying, but authentic communication is based on honesty and integrity. It is fundamental to authenticity that your audience believes that you will not outright lie to them. Engineering managers may not always be able to tell the whole truth in every situation, but conducting themselves with integrity is crucial. Most often, you can anticipate that you will be believed unless your actions or behaviors give people a reason not to believe you.

On top of truthfulness, authenticity comes from communications being perceived as unadulterated. We view others as more authentic when we believe they give us information straight without euphemisms or sugarcoating. As engineering managers, one of the best ways to reflect this in our communications is to acknowledge the difficulties that everyone already knows and understands. Rather than glossing over or ignoring failures, shortcomings, or bad situations, we can be viewed as authentic by addressing them head-on, particularly if they were caused by us—for example, stating plainly, "*We have this project coming up and we all know it's going to be painful, but let's just get through it together and it will be behind us in a month.*" Find the balance between when it's appropriate to communicate more artfully and when it's a better idea to have more rawness.

Showing your personality is another way to be seen as more authentic. People tend to be a bit more reserved in the workplace, but showing personality helps humanize us and makes us seem more genuine. Without personality, we may come across as too polished, robotic, and glib. Show personality by being yourself and letting some of your natural character and charisma come out. Use your authentic manner of speaking, your honest reactions to things, or even how you dress. Showing your unique personality allows people to feel that they are connecting with you and seeing you for who you are.

Lastly, be authentic by exhibiting vulnerability. As engineering managers, we spend much of our time projecting a brave face, but from time to time, it is a good idea to humanize yourself by sharing your fears or mistakes. It is core to your job to project confidence, giving your team confidence in you and themselves, but conveying a little vulnerability balances this out and has numerous benefits. Vulnerability helps bond the team by building trust and it lends authenticity by being sincere. So, don't be afraid of a little vulnerability at the right moment.

In the same vein as conducting our communications with authenticity, we can give authentic feedback using radical candor.

Giving feedback with radical candor

Radical candor is a communication technique based on caring deeply and challenging directly. Kim Scott's book *Radical Candor* introduces a communication quadrant with axes measuring from silence to direct challenge and from not caring to caring deeply. Depending on your leadership style and willingness to have difficult conversations, you may end up with one of four outcomes in the quadrant—radical candor, obnoxious aggression, manipulative insincerity, or ruinous empathy:

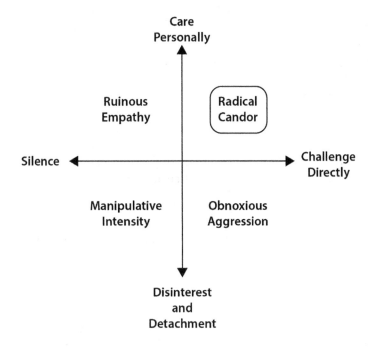

Figure 8.1 – Adapted from Kim Scott's Radical Candor framework

As an engineering manager who is invested in the success of your team and business, you probably care deeply without needing to be told to do so. We want success and growth for our engineers and coworkers. Challenging directly is what managers often struggle with. If you do not challenge directly (and do so with tact and caring) you will miss opportunities to help the growth of those around you. Growth demands honest feedback, learning, and course correction. Contribute to this virtuous cycle with radical candor in your communication.

Call on your bravery to face difficult conversations you need to have. When giving difficult feedback, be specific about what is expected, what was observed, and the outcome or impact of the situation. For example, if someone on your team is exhibiting problematic behavior, you can sit down with them in a one-on-one meeting and tell them, "*I expect that we have a professional environment where everyone on the team should be able to feel comfortable at work, so when you said _____ in the team meeting today, I was surprised. You may not have meant anything by it, but it was an uncomfortable moment for everyone. I want you to be successful, so in the future, I would like you to _____ instead. Does that sound reasonable and like something you can do?*" Notice how this statement delivers radical candor, expectation setting, and assuming the best by expressing surprise at the behavior. This is a direct correction that reinforces your confidence in the person.

With that, we have a broad understanding of what we want our communications to accomplish, so let's discuss how to best compose and structure them to be received and understood.

How to structure your communication

As engineering managers and team leaders, our communications may take many different forms. We may divide our time across informal chats, ad hoc updates, emails, structured written updates, meetings, presentations, and other channels of communication. Since we spend so much time acting as information conduits between individuals and groups, our goal is to make our communications as efficient and effective as possible. We want to deliver the right information in the right way so that we can keep everyone moving forward and avoid confusion, mistakes, and misunderstandings.

As we learned in the previous section, the best way to deliver information is with an audience focus. With that as our goal, let's go through the steps of how to think about structuring our communications. To determine the right approach for an audience, we will think about the format, duration, depth, and urgency.

Format

The format or method of communication is what medium the information is delivered in. Depending on the needs of the audience, it may make more sense to deliver a communication verbally or in writing and with different levels of formality and visual representation. Use the context and desired durability to decide on the best format. Ask yourself the following questions:

- Is this communication sensitive/private or public?

- Is this an announcement I want to capture in a shared team moment?

- Do I want to deliver this with emotion or is it more fact-based?

- Do I want people to be able to refer back to my message word for word?

- Is this better shared formally, informally, or both?

- Is this information more digestible in a visual format, such as a chart?

Verbal communication makes it easier to capture emotion or make an announcement widely. It is also more appropriate for personal matters delivered in one-on-one meetings. Written communication is asynchronous, so it's less disruptive and also makes it easier to refer back to the information exactly how it was delivered. Consider these trade-offs for your audience when deciding on the right format.

Duration

Duration refers to the length of the communication or the time it takes to consume it. For different audience needs, shorter or longer communication may be more appropriate and effective.

People often appreciate brevity in communication. If you can convey your message succinctly, that tends to be well received. Different audiences will have different expectations of this. For example, executive audiences often need the shortest, most to-the-point communications, since they have so many demands on their time. Consider your audience's point of view to determine whether a short update is best for communication.

In some cases, longer updates are more appropriate to provide more background, context, or opportunity for dialog. Longer communications have more opportunities to convey details and justifications. They can also soften communications and help you avoid coming across as rash or callous, which may happen with short communications. For these reasons, unpopular news is sometimes better conveyed in a longer communication.

You may also gain the benefits of both short and long communications by combining them. It often works well to start a communication with a short version and then transition into a longer version with more detail and explanation. This allows you to serve multiple audience needs by providing the key takeaways immediately and allowing the audience to hear more if they choose to. You can indicate this in your communication with headings or transitional phrasing, such as, "*Feel free to step out if that is all you need to know, but if you're interested, I will now go through the complete problem statement, options considered, and implementation.*"

Depth

The depth of communication is its granularity and framing. In other words, it captures how specific you are in communicating. An audience's level of understanding of a topic determines the appropriate depth of the communication.

For communications to be received and understood, the goal is to provide enough depth but not more than the audience can digest. Without enough depth, communications may not be believed or comprehended. With too much depth, communications may leave your audience completely lost.

As engineering managers, we generally know how to think about communications with full depth, but we may need to develop additional skills to pull that depth back to different levels for our audiences. Framing communications so that our audiences can understand them as much as possible is beneficial to working cross-functionally and with organizational leaders. Make use of examples, metaphors, and similes to help relate messages to different audiences. The more you can help them understand your work, the better your partnerships and team outcomes will be. Give your audience as much depth as needed while developing your ability to deliver messages in ways that make sense to them and keep them informed.

Urgency

The urgency of communication describes its level of immediacy to the audience. Depending on the nature of the communication, it may be best conveyed as soon as it is available or it may make sense to wait and provide it in the normal course of other information, such as in a periodic team meeting or a summary update.

Many communications are intuitively urgent or not urgent. For example, updates on service interruptions are obviously urgent. Nonetheless, it can be a good idea to specifically outline what types of communications you see as urgent and which you do not. You can confirm your assessment with your manager and senior leadership to make sure you are aligned on how you intend to conduct ongoing communications. This is helpful guidance to provide to your engineering team.

Once you have all of this information, you can put together a communication plan outlining who your audiences are and the appropriate format, duration, depth, and urgency for them. These plans are likely to evolve as you get to know your audiences and their needs better.

Now that we understand how to structure our communications, let's dive deeper into how we can approach communicating with specific audiences.

Communicating with your engineering team

Building upon the communication principles from earlier in this chapter, engineering teams have interests and priorities. Just like other audiences, we want to have empathy for these needs and tailor our communications to them, as well as determine what we believe they need to know to do their work and grow as engineers. To do this, let's go through the key communication touchpoints with our teams: one-on-one meetings, group meetings, and personal commitments.

One-on-one meetings

One-on-one meetings are standard practice for convening individually with a member of your engineering team or with others. These meetings are most often held at regular intervals as a means of catching up and privately exchanging information and feedback. They are typically scheduled for around 30 to 60 minutes and may or may not include project updates, along with general discussion.

When schedules are busy and teams are large, it can sometimes be difficult to find the time to hold one-on-one meetings with every member of your team. If you don't find the time to have these meetings, distance may grow between you and your engineers and you will have a harder time knowing what is going on with them. They may also think they are not important to you if you appear not to have time for them or if you have a habit of moving your meetings around frequently. Even when you are busy, have these meetings at least occasionally (such as once a month), respect the timeslot they are scheduled for, and remain completely focused on the conversation during that time.

For engineering managers, one-on-ones are an opportunity to connect with your engineers and make yourself available directly to them. By connecting with your engineers, you strengthen your relationship, build trust, humanize yourself, increase alignment, and develop a better understanding of each other. Having a strong connection increases comradery and loyalty and decreases the likelihood of attrition. By making yourself available, you can listen to them and potentially learn something important, demonstrate that you care about their well-being and perspective, and establish yourself as a manager that is open and approachable. These actions are good for your team relationships and help create an atmosphere of psychological safety.

One-on-one meetings with your engineers should have a balance between discussing projects, individual growth, and feedback exchange. Have an agenda in mind. You may or may not include status updates in your one-on-ones, but if you do, avoid making the one-on-one only a status meeting. Don't miss the opportunity to connect and make yourself available. Let's consider a few questions that can help frame a dialog with your engineers.

Projects:

- How is it going with your deliverables this week?
- How do you feel about the _____ project? Does it still seem achievable?

Individual growth:

- Do you feel like you are getting enough learning opportunities?
- How are you tracking against your career goals lately?

Feedback exchange:

- How are you feeling this week?
- How is your workload these days?
- What are your biggest challenges right now?

Other great resources for one-on-one questions include Lara Hogan's blog as listed in the *Further reading* section.

Some questions can be asked frequently while others about bigger career topics are best left to once a quarter or so. Tailor your questions to your engineers' careers and your workplace context.

Group communications

In addition to your one-on-ones and project meetings, it probably makes sense to have some group forums for your engineering team, either as meetings or written communications. Group communications for engineering teams are an opportunity to provide discussion, alignment, context, or updates. These can be established as ad hoc or periodic communications, such as an engineering sync-up every month.

Group communications can save time for you and can help bond together the engineering team. Communications better suited to the team environment include those for which everyone should receive the same information and those that may benefit from a communal setting. Examples of shared information communications include vision setting, leadership updates, announcements, training, and resource sharing. Examples of communal activities include gathering feedback on new ideas, problem-solving, and ideation.

For written communications with groups, in addition to audience focus you should aim to provide a digestible structure, avoid making assumptions about what people already know, and don't be afraid to overcommunicate by sending reminders. Structure your longer communications with headings, lists, tables, or anything else that makes them easier to understand at a glance. For example, if you are sending an email with two key concepts, make that immediately visually apparent with two headings. Define and explain as much as possible in your communications—in other words, write with your most junior team member in mind. Keep in mind that not everyone has your level of understanding. If your communication is requesting a response or that some action be taken by the recipients, make sure that is highlighted in the communication and send a reminder if you can. Remember, these are human beings with many other things on their minds. Reminders or other facilitation to make responding easier can help produce the response you are looking for.

For group meetings, be prepared and stick to the plan. Great meetings start with preparation. Make sure you have clear objectives for the meeting and that you have all of the right people. Before the meeting, write a memo outlining the material you intend to cover and any objectives you intend to accomplish. Ask the attendees to read the memo in advance of the meeting. With this preparation, you can focus on the tasks at hand and be respectful of everyone's time. During the meeting, it is your role to keep the group on task. If you have a group that is prone to distractions, tangents, or long-winded responses, gently guide them back to where you need to be to finish the objectives. For example, you may say, "*I think that's a good point, but it's not why we are here today, so let's table that for a separate discussion.*"

Personal commitments

Treat commitments you make to your engineering team with care. Personal commitments are a particular kind of communication you will have frequently with your engineers. Recognize that every commitment you make to your team is under scrutiny. Failure to keep even very small commitments harms your relationships and reputation.

As engineering managers, we are asked for things constantly. It may be for help, permission, consideration, advice, feedback, information, or support. With so many requests coming in, it can be easy for some to fall through the cracks and escape your attention. Your may be asked by one of your engineers in passing, "*Hey, could you take a look at this when you get a chance?*" If you say yes, but then fail to remember to follow through on the commitment, that may be a big disappointment for the person who asked you, who is then sat in a holding pattern waiting to hear back. You may figure that if it is important, they will remind you, but asking your manager for help can be hard to do already, so if they don't get a response, they are just as likely to think to themselves, "*I guess my manager is too busy to help me,*" and move on. These little disappointments add up to cracks in the foundation of relationships with your engineering team.

Avoid simple mistakes and forgetfulness by making it a point to keep a running list of everything you have committed to. Find a system for this that works for you. If you are often on the go, you can use a voice assistant to log commitments or you may jot them down in a notepad. Get in the habit of immediately cataloging every commitment you make.

Just as important as communication with our teams, communication with executives and senior leadership is an opportunity to help our engineering teams through advocacy. Let's take a look at how to do that.

Communicating with your leadership team

When we think of communicating upward to the senior leaders in our companies, we often focus on providing timely and accurate progress updates. While these updates are important day-to-day information, there is more to effective leadership communication. In addition to awareness of our progress, we need our leadership to be aware of our challenges and our value beyond particular deliverables.

Since executives and senior leaders have many demands on their time, they are interested in getting to the gist of communications. Our direct managers often have more appetite for and ability to digest the nuances of our day-to-day work, but with busy executives and senior company leadership, we can communicate effectively by focusing on the story of our team and its broader value, and reducing uncertainty for them.

Telling a story

Storytelling is a powerful means of communication. As humans, our brains are wired to seek and follow plausible storylines. This is a particularly effective means of conveying to executives what they need to know about your team.

For engineers, the story format can seem reductive and too convenient. As builders, we understand that reality is rarely a linear progression and true stories have nuance, idiosyncrasies, and general messiness. With our level of understanding, we may instinctively try to convey this depth to our leadership audiences, but getting into the weeds with them is usually not an effective means of contextualizing our work to those above us. We need to take an audience focus and distill our reality into a more digestible narrative format.

Rather than jumping in freeform, put together a structured narrative. The **CAR format** is where you focus on outlining the *context, action*, and *result* of a situation. The context is your scenario or problem, the action describes what you are doing to address the situation, and the result is the outcome—for example, *We have been struggling to fill the open roles on the engineering team because of competition with other employers in the bay area, so I have opened up the candidate pool to remote hires. This has tripled the candidates we have in our hiring pipeline and I am confident that we will be able to fill the positions before the end of the quarter*. Following this format helps you get out of your head and describe something more clearly from an outside perspective. Starting with the format allows you to reframe the communication from something complex into something simple and easy to follow.

Start with the CAR format and build upon it to develop your ability to tell stories that convey your vision, challenges, and how senior leaders can help you.

Conveying broader value

Any time you are giving a progress update to a leadership team, you can take the opportunity to convey why that work is important. As the engineering manager, you are keeping track of not just the progress on deliverables, but also the progress of your teams' skills, your delivery capability, your technology investments, and your engineering goals. That information is not only relevant within the confines of the engineering department but also to the business and leadership unit. If your leadership does not know about these aspects of progress, they will not understand the investments that have been made in technology and human capital. You are in a position to make sure they have this awareness.

First, use your storytelling techniques to frame each aspect of the work your team is doing in terms that can be understood by a leadership audience. Get your leadership team excited about even the most technical aims you are trying to achieve. There is always a business reason to be described that can contextualize your work for this audience. This might be enabling the team to move faster, reducing costs, increasing flexibility, creating new opportunities for innovation or insights, or something else. Metrics are particularly useful in conveying trends and improvements your team is responsible for driving.

Secondly, develop a sense of what your leadership team or teams care most about and further tailor your communications to those interests. It is beneficial to spend more time communicating around those areas or contextualizing your work concerning them. You can determine and communicate how each of your initiatives is driving or contributing to those areas. Use this to build executive support for your team's work and help leaders understand its value.

Reducing uncertainty

In your communications with senior leadership, always aim to reduce uncertainty. Executives spend their days making judgment calls and gathering information to inform those judgment calls. When you reduce uncertainty for them, you help them to make better judgment calls. Reduce uncertainty by providing simple, structured answers, giving a clear path forward, and finding a balance between directness, resoluteness, and confidence.

First, develop the ability to give simple, structured answers to questions. Storytelling with the CAR format is one form of structured communication. To reduce uncertainty, it is important to avoid giving muddled, ambiguous, or meandering answers. If there is nuance, focus on giving clear differentiators and decision points. Describe distinct steps or chains of events that provide useful context.

Second, present your path forward. Senior leaders want to see that you have considered potential obstacles and the best way to overcome them. There can still be unknown outcomes or unclear events in your path. But you can define decision milestones and points at which you anticipate changing course if needed. For example, you may say, *"I don't know whether this feature is feasible, but given the potential benefits, it makes sense to build a proof-of-concept for four weeks and assess at that point. If we have solved the design problems, we will continue; otherwise, we will be able to pivot to a fallback design approach."*

Lastly, reduce uncertainty in your communications by finding a balance between directness, resoluteness, and confidence. Be direct for timeliness and to build trust in what you have to say. Be resolute in your statements to show you have determination and conviction. Be confident to instill confidence in others. Together, these behaviors give certainty to your leadership team by showing them that you have a point of view and can follow through on your plans and handle obstacles.

Summary

In this chapter, you learned about the principles and techniques for great communication as an engineering manager. Communication is a critical skill and force multiplier to set teams up for success and produce good outcomes. Remember these tips when preparing any communication:

- Prepare those around you for what may come by setting expectations. Manage expectations with proactive communication at all times.

- Assume the best of others to avoid offense, promote trust, and encourage the right behavior. Don't jump to negative conclusions.

- Say no with care to produce acceptance and understanding. Take a structured approach to understand the request and alternatives, and decide what you can or cannot do.

- Adopt the perspective of your audience to help your communications be understood and remembered. Consider who your audience is and what is relevant and digestible to them.

- Maintain authenticity in your communications by acting with integrity, limiting euphemisms, showing your personality, and being vulnerable.

- Communicate feedback with radical candor by caring deeply and challenging directly.

- When planning your communications, structure them by thinking through the best format, duration, depth, and level of urgency.

- With your engineers, hold one-on-one meetings regularly, consistently at the same time, and inclusive of projects, individual goals, and feedback.

- Aim for group communications with your team that are focused, organized, and respectful of everyone's time.

- Track all commitments to your team, even the smallest ones, and be sure to uphold them.

- With executives and senior leadership teams, use storytelling formats such as CAR to keep your communications clear and engaging.

- Convey the broad value and business justification to senior leaders for every aspect of your work.

- Reduce uncertainty with your communications to senior leaders to garner their support for you and your team.

Now that you understand how to communicate effectively, you can use these skills to facilitate the process of evaluating and improving your engineering team in *Chapter 9*.

Further reading

- *Scott, Kim. Radical Candor (Fully Revised and Updated Edition): Be a Kick-Ass Boss without Losing Your Humanity.* St. Martin's Press, 2019. (`https://www.radicalcandor.com/the-book/`)

- Lara Hogan's one-on-one resources (`https://larahogan.me/resources/one-on-ones/`)

- Plucky's card deck for one-on-ones (`https://shop.beplucky.com/products/the-plucky-1-1-starter-pack`)

- *Stanier, Michael. How to Work with Almost Anyone: Five Questions for Building the Best Possible Relationships.* Page Two, 2023. (`https://www.mbs.works/how-to-work-with-almost-anyone/`)

- *Alda, Alan. If I Understood You, Would I Have This Look on My Face? My Adventures in the Art and Science of Relating and Communicating.* Random House Trade Paperbacks, 2018. (`https://www.penguinrandomhouse.com/books/533869/if-i-understood-you-would-i-have-this-look-on-my-face-by-alan-alda/`)

9

Assessing and Improving Team Performance

In addition to work on deliverables and immediate needs, engineering managers are charged with owning and improving team performance over time. Successful engineering managers must not only lead their teams but also effectively course-correct them when they are off track. Teams are complex organisms, so it may not always be obvious how to address their performance.

How do you know how your team is doing? And with that, how do you know what actions to take to optimize your team? These questions have fueled writings, research, and an entire industry of software and tooling that attempts to give engineering managers and business leaders some means to assess and improve engineering teams. In this chapter, we will survey the major approaches and research findings to understand how to solve these problems and develop a basis to hone your skills further.

This chapter takes the concepts you have learned in previous chapters and brings them together to accomplish the ultimate goal of all engineering managers: building great teams. We will introduce the most common approaches to understanding team dynamics, along with new research findings. You will learn how to assess engineers and the common pitfalls in doing so. You will learn how to determine what you are optimizing for and what factors to consider. We will introduce techniques for motivating, mentoring, and coaching to improve teams. By the end of this chapter, you will have a firm understanding of what to prioritize, why, and how.

This chapter is organized into the following sections:

- The classic stages of a team
- Assessing engineering teams
- Introducing team emergent states
- Improving team performance

We have a lot to cover in this chapter, so let's get started!

The classic stages of a team

Engineering ability alone does not determine the performance of a team. **Team dynamics** is a broad term to describe the collective behaviors and psychological processes that occur within a team. Engineering managers must explore concepts within team dynamics in order to understand why team performance is the way it is.

One of the most commonly used ways of looking at teams is through Bruce Tuckman's **four stages of team development**: *forming*, *storming*, *norming*, and *performing*. Developed in the 1960s, understanding these stages can give you a frame of reference as to why a team may be behaving in a certain way. Even without a deep knowledge of these stages, it is useful to be aware that a major contributor to team performance is where they fall on the path to team familiarity and acceptance. Keep this in mind as you work with and assess teams of your own. Let's look at the four stages in more detail:

- **Forming** teams are recently assembled and adjusting to being on a team together. They are typically experiencing a mix of excitement and anxiety, potentially creating some hesitancy in their tasks. Teams need clear structure, direction, roles, and trust-building to progress from here.

- **Storming** teams have moved past initial excitement and are experiencing frustration with the aspects of process or progress that they are not satisfied with, potentially arguing or becoming critical. Teams need help breaking down goals, receiving training, working through compromises, and managing conflict to progress from here.

- **Norming** teams are starting to get a handle on how to work together productively and resolve discrepancies in expectations. They are moving toward greater acceptance of others and feeling like a team. Teams need help evaluating processes and their productivity in a constructive manner.

- **Performing** teams feel satisfaction in the team's progress, acceptance of team membership, attachment to the team, and confidence in productivity. Maintaining your team at this stage is ideal.

Where do you think your team fits into these stages? Teams may get stuck at one stage or cycle between stages in response to change. Think about supporting your team to facilitate progression in their development stage.

Before we move on to team improvement, let's learn more ways to assess our engineering teams.

Assessing engineering teams

As engineering managers, we need to be able to accurately and consistently assess our teams. This is necessary to support and grow our engineering team's output, but also critically important to be prepared for times when the team's performance comes into question by others. When your stakeholders or company leadership question your team's performance, having a deep understanding of your performance trends and the data to support that will make your job much easier. Instead of reacting with uncertainty, maintain a working knowledge of team performance.

Assessing engineering teams comprehensively is hard to do. We may have gut feelings, the opinions of outside stakeholders, or raw measures such as lines of code written, but each of these can turn out to be poor indicators of team performance. In the search for an understanding of team performance, it is easy to go down the wrong path and arrive at the wrong conclusions. To avoid this, we will start by learning about the common pitfalls of assessing teams and then move on to specific measures, defining success, and assessing your team.

Common pitfalls of assessing teams

We may approach team performance assessment with the best intentions and common sense to guide us, but for the best outcome, we need a broad understanding of where this is known to fail. Notable performance assessment pitfalls include Goodhart's law, perverse incentives, and the McNamara fallacy.

Goodhart's law

We often want to support our decisions with quantitative data that we can measure transparently and monitor for trends over time. This good intention can sometimes be thwarted because of a concept known as **Goodhart's law**:

"Any observed statistical regularity will tend to collapse once pressure is placed on it for control purposes."

In other words, "when a measure becomes a target, it ceases to be a good measure" (DOI: 10.1002/(SICI)1234-981X(199707)5:3<305::AID-EURO184>3.0.CO;2-4). This economic concept has been broadly observed across statistics. It means that when we attempt to manage complex systems by reducing them to metrics, we end up optimizing for those metrics rather than our original complex goals. For our purposes as engineering managers, we should take particular care in how strongly we hold targets, since they may become less useful in response to organizational pressures. Metrics are still valuable as targets, but we should be wary and attentive to the changes that may result from our attempts to control them.

This also hints at the importance of selecting performance indicators with care so that you don't end up optimizing for the wrong behaviors and outcomes, as described next.

Perverse incentives

Sometimes when we create a target, it has unintended consequences. When we unintentionally incentivize the wrong behavior, this is known as a **perverse incentive**. For example, when building the first transcontinental railroad in the 1860s, the US government paid the builders per mile of track laid, leading the builders to deliberately lengthen the path and add unnecessary miles to the route. On software development teams, you can imagine similar problems may arise from naïve performance targets such as lines of code written or the number of tickets closed. When choosing performance measures and targets, avoid perverse incentives by considering extreme cases and the potential to game the system.

In addition to choosing performance indicators with care, we will want to avoid overreliance on our targets, as described next.

McNamara fallacy

When choosing and monitoring quantitative performance measures for our teams, we may become overly focused on a single target. When we quantify success on an inadequate set of measures ignoring qualitative data, this is known as the **McNamara fallacy**. In other words, performance cannot be reduced solely to convenient quantitative measures. Trying to reduce success to quantitative measures while ignoring all others is rarely a winning strategy. On software development teams, this could be something such as optimizing only for delivery speed and focusing on getting products into production as the measure of success while ignoring other factors such as customer satisfaction, team morale, and staff attrition. Such a measure is unlikely to produce stable team performance over time.

It is much more robust and insightful to incorporate both quantitative and qualitative measures in our performance assessments. With that, we understand the pitfalls to avoid, so let's examine these types of measures in the next section.

Quantitative and qualitative measures

All research relies on collecting and interpreting data. **Quantitative data** is inherently based on numbers and easily measured, such as build time. **Qualitative data** is non-numerical and relies on personal accounts, such as customer satisfaction. On engineering teams, quantitative data may be gathered through instrumentation or direct measurement. Qualitative data may be gathered through interviews, focus groups, questionnaires, or first-hand observation. Qualitative data collection involves more interpretation.

While both of these types of data have their limitations, together they can help engineering managers have greater insight into the performance of their teams and how they might improve. The question still remains: *What are the right things for engineering managers to measure?* Let's begin by introducing useful measures that we might decide to choose as performance targets.

Quantitative measures

DevOps Research and Assessment's (DORA's) *State of DevOps* is an annual survey that has gathered data on software teams' performance across the industry for more than a decade. In *Forsgren, Humble,* and *Kim's Accelerate* (*IT Revolution Press, 2018*), they identified the following quantitative measures as key indicators of engineering team performance and health:

- **Lead time:** The measure of how long it takes to go from the first commit into production deployment.
- **Deploy frequency:** Number of deployments in a given time, indicating how close a team is to rapid iteration. This is desirable in the vast majority of engineering team contexts.

- **Mean time to restore**: The measure of how much time it takes for a team to roll back or recover from a failure.

- **Change failure rate**: Percentage of deployments causing a failure in production.

In addition to DORA's measures, you can consider the following:

- **Team performance against specific service-level indicators (SLIs)**: In *Chapter 6*, we introduced SLIs as a means of assessing systems performance. We can also use this as a quantitative indicator of team performance.

- **Collaboration measures**: **Pull request (PR)** review count, PR comment count, and time to first comment.

- **Code quality measures**: Commit size, complexity ratings, static analysis scores, and code churn.

- **Staff demographics**: Staff attrition/churn rate, length of tenure, diversity of staff, and other demographic information.

- **Business metrics** that measure the success of the overall product may be a useful performance indicator. Depending on the business, quantitative measures might include **daily active users (DAU)**, sales, signups, or other user measurements.

Qualitative measures

Qualitative measures of engineering team performance seek to answer how the team is doing from various self-reported perspectives. One of the most widely used methods of qualitative assessment is **Net Promoter Score (NPS)**, which asks users to rate the likelihood that they would recommend something to a friend or colleague and then plots the response along a scale from *detractor* to *promoter*. This marketing assessment method is a simple way to estimate overall satisfaction.

We can consider the following qualitative data sources:

- **Self-assessment**: How do the team think they are performing? What do they think could be improved? What is their satisfaction level? This information may be gathered through questionnaires, interviews, or group sessions.

- **Stakeholder assessment**: How do stakeholders in the business view the team's performance? This might assess NPS, team strengths, team weaknesses, impact, and innovation gathered through questionnaires or interviews.

- **Engineering manager assessment**: Your direct observation can help assess more nuanced factors, such as impact and innovation.

- **Business metrics** may also be qualitative measures, such as NPS and other user-reported feedback.

Each of these data points can provide different insights and useful perspectives on the performance of your team. Together, they allow you to observe patterns and come up with hypotheses for improving the performance of your team.

Before you assess your team, you must determine what your basis of assessment is. In other words, you need to recognize what success looks like and what specifically you are optimizing for.

What is your success definition?

As with most things we have discussed in this book, optimizing team performance is not a *one-size-fits-all* pursuit. Team performance is contextual to the priorities of the business, the style of work, and the product itself. A team that is very effective in a particular context might perform poorly if switched to a very different context and set of expectations. To assess your team, we will begin by assessing the context in which your team works and the needs of that context. To do so, we will consider *efficiency*, *effectiveness*, and *innovation*.

Efficiency means accomplishing a goal with the minimum amount of resources, such as time, staff, or raw materials. Something is considered efficient if it functions with only those resources absolutely needed. Effectiveness means producing the intended result. Something is considered effective if it is accurate and precise in achieving the end goal. Innovation in this context means discovering a new solution to a problem. Something is innovative if it proposes an entirely new way of achieving a goal, either from a product or technology perspective. While almost every team strives for efficiency, effectiveness, and innovation, their relative importance is very different at different stages of growth or depending on the business goals. Mature products tend to prioritize efficiency, experimental products favor effectiveness, and innovation is usually the highest priority only in the research domain. The table in *Figure 9.1* shows success definitions across these software engineering contexts.

Setting	Primary goal/horizon	Guiding questions	Success definition
Existing mature product	Improve	Are we maximizing our resources in delivering this proven product?	Efficiency
Emerging and experimental	Expand	Is the product good? Does it do what we intend?	Effectiveness
Research	Disrupt	How can we produce exponential improvement in the existing market?	Innovation

Figure 9.1 – Success definitions

This table can help simplify and differentiate appropriate optimization targets across contexts. For example, most early-stage start-up ventures are experimenting with a new product; they have a problem domain and a product solution to test, so the success definition is effectiveness.

If you find that the table doesn't suit your situation, try asking yourself the following questions:

- Is there more emphasis on delivery consistency or delivery accuracy?

- Are we working toward incremental improvements or exponential change?

Use your knowledge of the business, departmental goals, feedback from your leadership, and common sense to determine the right success definition. Now that you have a basis for assessment, you can move on to assessing your team.

Assessing your team

Teams are best assessed in their entirety because they produce outcomes from joint efforts. Use your success definition to guide you in determining the ideal performance target measures for your team. We identified at least 10 useful measures to consider earlier in the chapter, so let's revisit them and how you might interpret each:

- **Lead time** is an efficiency metric since it measures the team's usage of time resources. It can be pulled from your issue tracker's issue duration, or instrumenting a combination of that and VCS reporting.

- **Deploy frequency** is an efficiency metric since it measures the ability to deliver small changes often and with high confidence due to automation. It can be pulled from your CI/CD pipeline reporting.

- **Mean time to restore** is an efficiency metric since it measures the maturity and sophistication of your versioning as reduced to a time factor. This can also be pulled from your CI/CD pipeline reporting.

- **Change failure rate** is an effectiveness metric since it concerns how effectively the team has produced a new version. A ratio measure of *bugs:version* demonstrates how effective a version is in solving the problems it intended to solve. This metric can be pulled from your issue tracker.

- **SLIs** on development teams are generally systems performance measures and thus effectiveness/innovation metrics. For example, common SLIs might be *API response time* or *time to interactivity* for a user interface. They concern the effectiveness of accomplishing what is intended and, potentially, innovation in measuring new ways of accomplishing goals using different methods. An innovation-focused team might work on the exponential improvement of systems performance in a particular area. These metrics are reported and gathered in the monitoring suite of tools.

Note

On purely maintenance- or operations-oriented teams, examples of SLIs are more likely to be customer-service oriented, such as issue duration or the time to acknowledge an issue. This sort of SLI is an efficiency metric.

- **Collaboration measures** such as PR reviews and comments tend to be good effectiveness/innovation metrics. Discourse and even conflict at the task level have been proven to produce superior creative solutions in a team setting. Increased discussion aids effectiveness and promotes innovation, though the tone and sentiment of the discussion moderate this and should be observed. These metrics can be pulled from your VCS reporting.

- **Code quality measures** such as commit size, complexity ratings, static analysis scores, and code churn are effectiveness metrics because they measure how well something is accomplished rather than how few resources are used. They can be collected from across your CI/CD pipeline tooling or specific quality instrumentation such as Code Climate or Codacy.

- **Staff demographic measures** vary in optimization relevance, with the most salient being the relationship between diversity and innovation. It has been repeatedly shown that all forms of diversity promote innovation by giving the team a broader base of experience and perspective to draw upon. These metrics can usually only be gathered through your HR department.

- **Business measures** (qualitative and quantitative), if deemed applicable, tend to be effectiveness/innovation metrics since they provide insight into end-product performance. They are usually collected via analytics reporting tools.

- Team **self-assessment** and **stakeholder assessment** are both neutral performance metrics since they concern personal interpretations, which may be oriented toward any success definition. If the team is highly aligned toward a consistent success definition, they may be more reliable. These metrics can be collected with questionnaires, interviews, or group sessions such as a sprint retrospective.

- **Engineering manager assessment** can be much more nuanced and focused on any success definition. Developing your own gut feeling helps to round out the big picture of performance. You can use the *guiding questions* from the table shown in *Figure 9.1*.

Given these measures and their focus areas, you can assemble a collection of team performance indicators that is ideal for your context and goals. As mentioned, all teams are concerned with efficiency, effectiveness, and innovation, but they are more concerned with one of the three. So, a good collection of performance measures will include each of these but more of one type.

For example, let's say you work at a small start-up company looking for product-market fit on an idea the founders have for a new auction web app. This is an emerging and experimental setting, so your highest priority is going to be effectiveness. You might write the following plan:

Team **key performance indicators (KPIs)**	Deploy frequency (efficiency)
	Change failure rate (effectiveness)
	Collaboration: # of PR reviews (effectiveness/innovation)
	Quality: static analysis score (effectiveness)
	Business: daily active users (effectiveness/innovation)
Also measured	Lead time (efficiency)
	Mean time to restore (efficiency)
	Systems performance SLIs (effectiveness)
	Staff demographics

Figure 9.2 – Example of team performance objectives

You can share the "KPIs" with the team and reserve the "also measured" metrics as additional detail for you to be aware of. There are some judgment calls made in this selection, such as deciding that, for a preferred efficiency metric, you will choose deploy frequency over the others since you are working on proving a new product, and you estimate that speed will be most relevant. You are using all of the effectiveness metrics with the exception of systems performance since ostensibly, users' willingness to use the product supersedes its raw performance as an effectiveness measure (and you expect this information to be included for the team in your monitoring dashboard). We can criticize these choices by asking whether the choice of collaboration metric could result in a perverse incentive by encouraging the volume of PR reviews without accounting for the depth or quality of those reviews. This would be something for you to pay attention to. You may also find that PR quality and quantity are happening but are slower than expected, suggesting you may want to incorporate a measure of PR duration, such as *time to merge*. You can—and should—adjust this over time as needed.

The exact target values for these measures are less important than observing how they trend over time. You can set some initial targets contextual to your team if you think that makes sense, such as saying you want every PR to receive at least two reviews. Use your judgment to decide the necessity of this, being careful to avoid setting up targets that seem unrealistic or insurmountable to the team.

With this, you now have a solid data-driven basis for how to set performance goals and assess your team, but you may be wondering: *What about influencing the more intangible aspects of team performance, such as engagement, cohesion, and psychological safety?* You probably know that these are important, but how do you develop them on your team? For this, we will build on our current understanding as we move into the next section and learn about team emergent states.

Introducing team emergent states

In the past few decades, researchers have sought to shed light on some of the underlying factors contributing to team performance and outcomes. Marks, Mathieu, and Zaccaro (2001) introduced a taxonomy for describing teamwork and **team processes** that identifies **team emergent states** along the way. Their definition of team process can be summarized as *interdependent acts to produce outcomes in taskwork and achieve collective goals*. Team emergent states are distinguished from processes as *dynamic and contextual states of being that are reflected in members' beliefs and behaviors.*

Further studies (DOI:10.1146/annurev-orgpsych-012218-015106) have suggested that the primary factors that predict performance and outcomes for teams include the following:

- **Compositional features**: Who is on the team—their skills, traits, and tenure

- **Structural features**: How work is structured—tasks, interdependence, and virtuality

- **Leadership behaviors**: How the team is led—style, participativeness, and openness

- **Processes**: Methods of converting work inputs to outputs

- **Emergent states**: Cognitive, affective, and motivational states of teams

We have hinted at team emergent states throughout this book, but now that our focus is on team performance, a deeper understanding of them is needed. For the rest of these performance factors, you can refer back to *Chapter 2* for a refresher on leadership styles and behaviors, *Chapter 5* for processes and structural features, or jump ahead to *Chapters 13* and *16* for more on team compositional features.

Team emergent states matter to engineering managers and teams because they are proven to have an impact on team performance and outcomes. They may be less approachable than other performance factors, but having an understanding of how and why these states occur can give you an advantage as a leader in your organization.

Since team emergent states are an abstract concept, let's go through the most common examples. Collectively, these are *trust, psychological safety, confidence, cohesion, cognition, team climate, identification,* and *empowerment* (DOI:10.1177/1046496420956715). As team emergent states, these are described as follows.

- **Trust**: Team members' beliefs in the dependability and trustworthiness of each other

- **Psychological safety**: Team members' beliefs that the team is safe for interpersonal risk taking

- **Confidence**: Team members' beliefs in the team's general and task-specific abilities

- **Cohesion**: Team members' willingness and likelihood to remain united in pursuit of goals

- **Cognition**: Consistency of team members' shared mental models and other knowledge organized, represented, and distributed within the team

- **Team climate**: Team members' perceptions of norms, attitudes, and expectations

- **Identification**: Team members' shared sense of identification as a workgroup
- **Empowerment**: Team members' beliefs regarding their authority and responsibility for their work

An interesting point about team emergent states from research is that they are simultaneously outcomes of actions and inputs to actions. An example of this is given by Marks, Mathieu, and Zaccaro (2001): *"teams with low cohesion may be less willing to manage existing conflict, which, in turn, may create additional conflict that lowers cohesion levels even further."* So, an input of low cohesion on a team may produce an output of lower cohesion. Observation and attention to these dynamics can help you understand behavioral trends on your team.

Now that you know what team emergent states are and why they matter, let's move on to the important part—how to prioritize and influence them.

Prioritizing desired team emergent states

At first glance, our commonly sought team emergent states seem universally desirable, but that may not always be the case. For example, cohesion sounds good, but have you ever been on a team that was so cohesive that team members cared more about supporting each other than producing the best outcome for the product? This could lead to a destructive result on a project that needs more proliferation and debate of ideas. In some settings, too much of an emergent state can be a bad thing.

Similar to the success definitions we introduced earlier in this chapter, we can consider that each of these team emergent states is valuable to all teams, but their relative importance is a key consideration for optimizing team performance. To avoid misalignment of emergent states, let's take a moment to examine which emergent states are most suitable to the success definitions and other contextual factors, as follows:

- **Trust** is particularly foundational to innovation. It is also critical for teams that have remote work structures (DOI:`10.1037/apl0000113`).
- **Psychological safety** is foundational to innovation since innovation requires taking risks.
- **Confidence** is most important when task interdependence on the team is high. The more team members rely on each other, the more important their confidence in the team is.
- **Cohesion** is most important for efficiency.
- **Cognition** is particularly important for efficiency and effectiveness.
- **Team climates** are particularly useful for promoting other emergent states and fine-tuning individual-level outcomes. So, for example, if your goal is innovation, you may want to develop a team climate with norms and expectations of psychological safety, trust, and valuing innovation.
- **Identification** paired with psychological safety is noted to have a strong effect on a team's ability to learn, suggesting that this could be an ideal combination for promoting effectiveness in teams.
- **Empowerment** is particularly important to effectiveness and innovation. It is also great for promoting individual positive outcomes such as engagement, job satisfaction, and commitment.

With these guidelines, you can decide which team emergent states are your top priorities and focus the majority of your efforts there. Choose two or three, but definitely not more than four of these so that you don't dilute your priorities. For example, if your team is working on an existing product at a mid-sized company, you might choose efficiency as your primary success definition. You might decide to focus on developing high cohesion and high cognition in support of that. If you consider that your team's work isn't very interdependent and so team confidence isn't the highest priority, but you want them to feel committed and engaged with their work, you could add high empowerment to your core list of desired team emergent states. Now, you have a few to focus on.

Let's go through another example of this. Consider you are working on product market fit at a start-up, so you choose effectiveness as your primary success definition. You might decide to prioritize high cognition, focusing on each engineer having the same understanding of the problem space. Then, you could prioritize high identification and psychological safety to promote the team's ability to learn from rapid iteration of the product. These emergent states make sense to boost your team's performance against your product goals and business metrics.

Once you have identified a few team emergent states to work on, you can move on to how to develop those states in the next section.

Fostering team emergent states

Developing team emergent states is complex and gradual. It is not something you will be able to accomplish overnight, but the result of consistent efforts and conscious leadership on your part.

Fortunately, there have been many studies on the precursors to team emergent states (DOI:10.1177/1046496420956715). Basic antecedents for our list of team emergent states include the following:

- **Trust** precursors are individual skill, integrity, and emotional intelligence. These precursors should be demonstrated by you and reinforced in team members.

- **Psychological safety** precursors are leadership behaviors, positive leader relations, trust, identification, and supportive contexts. Be open, non-judgmental, caring, and non-punishing to encourage this.

- **Confidence** precursors are organizational support and formalization, composition, shared/ transformational leadership, trust, and empowerment. Help engineers to know where they stand and to believe in the team's ability to achieve their objectives.

- **Cohesion** precursors are member affinity and perception of fair equitable treatment. Spend time developing relationships on the team and building a team climate of fairness.

- **Cognition** precursors are task interdependence, team design (role differentiation), participative leadership, and team debriefing processes. Invest time in discussing your product, systems, and the problem space deeply.

- **Team climate** precursors are leadership styles and behaviors, demonstrating and emphasizing those expectations. Shape team climates by saying, writing, doing, and reinforcing those norms and expectations.

- **Identification** precursors are leadership styles and behaviors. Provide guidance, encourage involvement, and lead by example.

- **Empowerment** precursors are leadership styles emphasizing participative decisions and organizational support. Demonstrate openness to and expectation of autonomy, and ensure team members have access to necessary organizational resources.

Use these precursors to create the conditions where your desired team emergent state can grow over time. For example, if your goal is high cohesion on the team, you could work on raising member affinity through consistent relationship-building and team-bonding activities. You would also pay careful attention to the perception of fairness and demonstrating inclusive and equitable behaviors. Alternatively, if you were unsure of how to increase the team's confidence, you might ask them individually, *What would it take for you to have more confidence in the team?*

Now that you are on the path to developing productive team emergent states in alignment with your chosen success definition, we can shift our focus to what other actions we may take to improve team performance.

Improving team performance

In this chapter, we have learned powerful techniques to focus your efforts and work toward ideal outcomes as an engineering manager. Along with your success definition, performance targets, and desired emergent states, you need ways to motivate, mentor, and coach your engineers.

Motivating your team

In *Chapter 6*, we introduced the importance of motivation in doing our best work, but only in the context of supporting production systems. Maintaining and encouraging motivation in your team is helpful in all aspects of their work, so let's look at motivation more broadly now.

If you believe that workers are inherently lazy, then you have what Douglas McGregor dubbed a *Theory X* management style. McGregor's hypothesis calls managers *Theory Y* if they make positive assumptions about their teams, believing they are genuinely interested in and committed to their work. *Theory X* managers make negative assumptions, viewing their team as self-serving and likely to do the least work possible. Engineers and other knowledge workers are demotivated by *Theory X*-style management, so it is good to be conscious of this when examining your beliefs about how to motivate your team.

In 2009, Daniel Pink published his book *Drive: The Surprising Truth About What Motivates Us*. In this seminal book, he outlined three key areas of impact for intrinsic motivation of knowledge workers: autonomy, mastery, and purpose. Engineering managers can use Pink's findings to help motivate and increase performance on their teams.

Autonomy

Our desire to be self-directed and responsible for our own outcomes is *autonomy*. An autonomous work environment provides team members with the agency to determine what to do or how to do it. Autonomy promotes engagement and encourages intellectual curiosity and learning.

Engineering managers can give their teams autonomy by adopting a shared-decision leadership approach. This means having participative decision meetings and/or delegating decisions directly to your team. In participative sessions, you can facilitate shared decisions by giving your engineers the context for the decision that needs to be made and then opening up a discussion of what ideas they have, discussing trade-offs together. Alternatively, you may choose to delegate to the team or a specific engineer. In the case of delegation, you can provide more or less autonomy by delegating in different ways. You might ask the team to come back to you with a proposed solution and justification, giving them a good amount of autonomy without too much risk. Or, you may say, "*This decision is yours and you should be prepared to support it and handle any consequences.*" Giving the maximum autonomy can produce the most motivation, but proceed with caution since it also introduces risk.

Mastery

Our desire to become skillful at our work is *mastery*. A workplace environment that emphasizes mastery rewards skillful contributions from team members. This environment encourages engineers to strive to be the best at their craft.

Engineering managers can support mastery in their workplace by demonstrating, recognizing, and rewarding skillful contributions and skill growth on their team. First, you can demonstrate mastery through your own behaviors, continually growing your skill set. Then, recognize skill on your team, doing so only for genuine skillful contributions and not for appearances or attempts. In other words, you will only encourage mastery by recognizing mastery; if you recognize all efforts the same, it won't be effective. Regular and expected rewards can actually diminish performance. Finally, further reward skill building with increased autonomy, position, and other perks that you come up with.

Purpose

Our desire to do meaningful and impactful work is *purpose*. A purposeful work environment gives work meaning and significance, connecting our day-to-day actions with a motivating cause or result. This environment gives engineers fuel to keep going since they believe their work is making an important difference.

Engineering managers can support a purposeful workplace by understanding and contextualizing the team's work. Think about this from top to bottom; try to convey meaning for the company, product, feature, component, bug, or line of text. If the purpose of your company is unclear, work with your leadership to understand it better and translate this to your team.

Consider how the work is making a difference and provide your team with that perspective. Humanize your end users—talk about their challenges and how your work helps them. If you are building a form input, for example, talk about how frustrating it can be when validation error messages are confusing or not user-friendly. Your form validation is an opportunity to give someone a happy moment by empathizing with them and designing the behavior in such a way that they have a great experience. Encourage your team to really think about that person, how important this may be to them, and how they may benefit from this. For example, imagine someone's grandfather is trying to sign up for your app to see their grandchild's posts and trying to get through the form with little technology experience. Finally, convey metrics that demonstrate the impact of the work. This type of thinking produces not only motivation but also better product outcomes.

Another way to convey purpose is by highlighting how the work helps peers or colleagues within the company. Your work may be impactful and meaningful to your stakeholders in other departments of the business. Connect with stakeholders to learn how your work helps them and bring that information back to your team.

Weaving autonomy, mastery, and purpose into our understanding of our work produces long-term satisfaction and motivation for our teams. Now that you have a handle on the basics of motivation, let's move on to mentoring and coaching.

Mentoring and coaching individuals on your team

Giving guidance and advice is core to improving the performance of software engineering teams. Guidance where we closely assess, prescribe, and direct is known as **coaching**. Guidance where we listen and make suggestions with more leeway is known as **mentoring**. Often, mentoring is self-directed, with the mentee bringing challenges to the mentor to discuss. Most times, coaching is more appropriate for team members with less experience, while mentoring can be better for those with more experience.

Whether you structure your guidance as mentoring or coaching, use these steps to help make it a productive session:

- As the mentor or coach, keep in mind the following points:

  - Self-awareness and self-honesty are hard, so you may need to be patient as these are built up over time through your guidance

  - Each individual is different, will have different challenges, and move at a different pace

- Prior to mentoring or coaching, make sure the timing is right to have a positive session. If they are tense and defensive (or you are), give them time to cool off.

- For mentoring, the best starting point is their goals, so begin by establishing what those are and why.

- For coaching, the starting point can be their role and workplace expectations. Use these as the basis for contextualization and feedback.

- To identify the needs of the individual, ask questions and listen intently. Are there expectation mismatches? Are they misreading situations? Are there specific skill gaps?

- If needed, use the **GROW** model to formalize a plan:

  - **Goal**—What do you want? (Or what is expected of you?)

  - **Reality**—Where are you now?

  - **Options**—What could you do?

  - **Will**—What will you do?

- For corrective feedback, include clear statements of what behavior was expected, what behavior was observed, and what the result of that behavior was.

- Destigmatize mistakes; make it clear that the aim is progress, not perfection.

- Leverage teachable moments with questions such as *Why do you think that happened?* and *What do you think you could have done to avoid that?*

- Use radical candor to provide honest and caring feedback.

- Get verbal confirmation of any behavior changes you ask for, such as asking *Do you think you can do that from now on?*

- Don't overwhelm them with too much information at one time.

So, for example, let's say there is a new engineer on your team who is submitting large PRs of their whole week's worth of work. Your expectation is that engineers will submit smaller PRs as they work so that they will be easier to review and easier to merge. You've identified this as an opportunity to use coaching to improve their work habits. You can start by pulling them aside to ask and listen. *"I saw you submitted a big PR of the whole week's work… how come you didn't break that up throughout the week like we usually do?"* With that question, you have re-established the expectation (small PRs) and asked why without jumping to any conclusions. At that point, you can listen for what their individual needs might be, leverage any teachable moments, and get verbal confirmation of behavior change.

Mentoring and coaching are great ways to help engineers on your team grow and reach their full potential. Practice these steps to become more natural at them.

Before we close the chapter, let's take a moment to talk about remote teams and how to help them be successful.

Improving remote teams

Teams whose individual members are geographically distributed are known as remote or virtual teams. Remote teams have become increasingly common in recent years. Remote engineering teams have lower office real-estate costs and are able to offer their engineers enticing perks such as working from the comfort of their home, having no commute time, no commute costs, and increased work/life balance. Remote teams are subject to the same leadership factors as colocated teams, but there are a few specific guidelines to help them be successful and productive, as follows:

- Trust between team members and leaders is even more important in remote teams. Since team members do not have visual cues and body language accessible to them, engineering managers can support engineers by prioritizing a trusting emergent state and team climate.

- Preparation is often more important for successful remote interactions and group sessions. Since everyone is in their own space, teams need proper tools and contextual information provided in advance to allow collaboration. Provide resources not just for communication, but also screen sharing, pair programming with multi-cursor tools (such as CoScreen, Pop, or Drovio), and whiteboarding (with tools such as FigJam or Miro).

- Asynchronous communication methods are especially well suited to geographically dispersed teams. Consider moving status updates to a written format in a group chat room. Adopting a memos-over-meetings working style can be beneficial in all settings, but especially so for remote teams that may work different hours.

- For teams that are partially remote and partially colocated, adopt a remote-first approach. When some staff are colocated in a conference room and some are remote, it is easy for those remote members to become deprioritized or overlooked. Adopt a remote-first approach by instead having everyone join meetings from their computers so that everyone is on equal footing.

- Find ways to have team bonding, informal socializing, and celebrations remotely. Ask your manager for celebrations funding and use it to mail your staff swag or other treats. Create space and time for this at least occasionally.

Together, these practices can help your team to identify as a team, feel valued, and be empowered to have a productive remote work day.

Summary

In this chapter, you learned how to assess and improve the performance of your engineering team. To be successful as engineering managers, we must be able to understand our team's performance and how to guide them to meet the challenges of our businesses. Building high-performing teams requires an understanding of measurement, goal setting, managing team dynamics, and working with individuals on your team.

Key takeaways include the following:

- Be aware of and help your team progress through the stages of teams: forming, storming, norming, and performing

- Avoid the pitfalls of assessing teams: Goodhart's law, perverse incentives, and the McNamara fallacy

- Determine what high performance means for your team by choosing a success definition: efficiency, effectiveness, or innovation

- Assess your team with both quantitative and qualitative measures to have a complete picture of their performance

- Choose contextual KPIs and share them with your engineering team

- Understand and foster productive team emergent states that support your performance goals

- Motivate your team by creating a working environment that provides autonomy, mastery, and purpose

- Use coaching and mentoring to address individual performance on your team

- Give your remote team a support structure, resources, and a working style that allows them to do their best work

Incorporating these techniques into your engineering management practice creates an environment that supports and encourages high performance. In the next chapter, we will build on this foundation to develop a highly accountable engineering team.

Further reading

- *Using the Stages of Team Development. MIT.* (https://hr.mit.edu/learning-topics/teams/articles/stages-development)

- *Forsgren, Nicole, Humble, Jez,* and *Gene Kim. Accelerate: The Science of Lean Software and DevOps: Building and Scaling High Performing Technology Organizations. IT Revolution, 2018.* (https://itrevolution.com/product/accelerate/)

- *DORA Quick Check* (https://dora.dev/quickcheck/)

- DORA publications, resources, and community (https://dora.dev/)

- *The SPACE of Developer Productivity. ACM Queue, 2021.* (https://queue.acm.org/detail.cfm?id=3454124)

- *Pink, Daniel H. Drive: The Surprising Truth About What Motivates Us. Penguin, 2011.* (https://www.penguinrandomhouse.com/books/301674/drive-by-daniel-h-pink/)

- *How to mentor software engineers. xdg.me, 2021.* (https://xdg.me/mentor-engineers/)

- *How to Give Feedback to People Who Cry, Yell, or Get Defensive. hbr.org, 2016.* (https://hbr.org/2016/09/how-to-give-feedback-to-people-who-cry-yell-or-get-defensive)

10
Fostering Accountability

The work of software engineers involves making individual contributions toward shared team goals and objectives. In making contributions, they are each given responsibility for some portion of those shared goals. Even so, we may not always be able to count on everyone to display the same level of responsibility and commitment to their work. How can engineering managers encourage their engineers to take an active role in team objectives?

To manage a high-performance software engineering team, you will eventually need to think about **accountability**. In a workplace setting, accountability is the willingness to take responsibility for one's actions and their outcomes. Accountable team members take ownership of their work, admit their mistakes, and are willing to hold each other accountable as peers. In performance terms, from *Chapter 9*, high accountability is a desirable **team climate**.

In writing and research on managing teams, accountability is often highlighted as a key factor in team performance. Why is that the case and what can engineering managers do to foster accountability in their teams?

In this chapter, you will learn why accountability is so often tied to team performance, what its benefits are, and your role in maintaining it. You will learn how to instrument accountability and what steps to take to produce an accountable team climate. We will go through examples of accountability in practice and how to handle specific situations. By the end of this chapter, you will have a robust set of tools for building highly accountable teams and solving accountability problems.

This chapter is organized into the following sections:

- Accountability and performance
- Building an accountable team culture

Let's start by gaining a deeper understanding of the relationship between accountability and performance.

Accountability and performance

In your career, you may have had experience with a low-accountability team. Low-accountability teams can be recognized based on their tendency to shift blame, avoid addressing issues within the team, and escalate most problems to their manager. In low-accountability teams, it is difficult to determine the root of problems, failures are met with apathy, and managers have to spend much of their time settling disputes and addressing performance. Members of low-accountability teams believe it is not their role to resolve disputes and instead shift that responsibility up to the manager, waiting for further direction. These teams fall into conflict and avoidance deadlocks, unable to move quickly because they cannot resolve issues within the team.

On the other hand, high-accountability teams are characterized by having members that are willing and able to resolve issues within the team. They take responsibility for their own actions and hold each other accountable. They take ownership of resolving disputes and feel empowered to do so without intervention from others. They learn quickly by identifying issues and solutions together, adopting better patterns over time. They are able to work without delay because they don't need anyone else to resolve problems. Their managers are able to work more strategically without being bogged down by day-to-day conflict resolution.

When team members take ownership of work problems and solutions, teams move faster because they are not waiting on someone else to deal with problems for them. Instead of funneling through a bottleneck decisionmaker, high-performance teams have a more efficient peer-to-peer approach to problem resolution.

In his article *The Best Teams Hold Themselves Accountable*, Joseph Grenny writes the following:

- *In the weakest teams, there is no accountability*

- *In mediocre teams, bosses are the source of accountability*

- *In high performance teams, peers manage the vast majority of performance problems with one another*

"…The role of the boss should not be to settle problems or constantly monitor your team, it should be to create a team culture where peers address concerns immediately, directly, and respectfully with each other."

Holding ourselves and each other accountable for our actions and outcomes creates an environment where teams not only move quickly, but also have high consensus on expectations, trust, and integrity. Highly accountable teams inspire confidence both inside and outside of the team, developing a reputation that promotes further trust and success.

Now that you understand the relationship between accountability and team performance, let's learn how to manage accountability.

Building an accountable team culture

When we look deeply at accountability as a trait of teams, we may break it down into two basic components: the acceptance of responsibility for outcomes and the willingness to personally fulfill that responsibility. In other words, accountability concerns the belief that the work is ours and the belief that we have the agency to carry it out. Our goal as engineering managers is to guide our teams to a state where they possess both of these beliefs.

Internalizing ownership and internalizing agency are two very different goals. You can imagine how easy it would be to know that a job is yours while having no idea how to do it. For team members to feel responsible for work but lack the ability or environment to accomplish that work can be incredibly demotivating and damaging. This underscores the importance of supporting both of these aspects of accountability on our teams. To serve these dual aims, use the three Ps of accountability: provide, promote, and practice.

Provide guidance
Promote ideal behaviors
Practice during crises

Figure 10.1 – The three Ps of accountability

Provide guidance for accountability by giving information and resources, promote accountability through reinforcement of behaviors, and practice accountability during high-visibility situations where all eyes are on you.

Providing guidance

To provide guidance for accountability is to ensure that your team members have everything that they need to successfully carry out the work. Providing guidance means that your team understands the work, its goals, and their own roles, and has access to necessary information and tools. Let's go through each of these.

Building a shared understanding of work

A shared understanding of work refers to your team's cognition or mutual grasp of their domain, systems, and processes. Does every member of the team have this awareness? A shared vision of your product and systems is the foundation for aligning your team's understanding of the work. We also introduced the idea of shared mental models in the previous chapter.

As you learned in *Chapter 9*, the precursors for shared cognition and mental models are the following:

- Task interdependence
- Team design and role differentiation

- Participative leadership
- Team debriefing processes

Since we will address role differentiation and leadership behaviors in further sections of this chapter, we can focus here on the remaining two ways to establish shared cognition on teams, task interdependence and team debriefing processes.

Develop shared mental models on your team by arranging task interdependence, positioning team members to work closely together to accomplish their work tasks. Prevent mental models from diverging by avoiding situations where work is done too independently for too long. Encourage crossover with mutual brainstorming and planning activities. Arrange projects and tasks with the aim of creating a shared understanding of the domain and how systems work. Practice pair programming regularly.

Secondly, establish group debriefing processes to keep everyone on the same page. Aim to give everyone on your team the same level of context so that they all can develop the same grasp of products, features, and systems. Favor group update sessions over individual updates. Make debriefing a consistent part of your team routine and incorporate it into your onboarding process. Try capturing this information as cataloged memos or video recordings to allow you to share it as needed.

If you can do this consistently for the business, domain, systems, and processes, your team will have a strong shared mental model to work from that helps them to make good decisions and resolve challenges. Along with providing a shared understanding, you'll want to be explicit about what success looks like.

Maintaining clear goals with success definitions

Provide clear goals by confirming you have sufficient success definitions in place for your team. Communicate unequivocally what is expected of them.

Similar to shared understanding, success definitions are another area where vision setting forms the basis. Your company, product, and engineering visions are high-level definitions of success for your team's work. In *Chapter 9*, we determined success definitions for engineering team performance. Make sure you have clear success definitions all the way down to individual units of work.

To provide for accountability, ask yourself the question, does everyone understand and agree on what success looks like? Is this true at the task level as well as at a high level? An example of success definitions at the task level is the acceptance criteria we introduced in *Chapter 5*. With acceptance criteria in place and used diligently, team members have an unambiguous picture of what they are working toward for each and every task.

You can take a similar approach to engineering expectations and tasks. Ask yourself, are code quality practices and conventions clearly defined? Utilizing explicit and unequivocal resources such as code style guides, linting, and pull-request templates provides success definitions that make engineering expectations both clear and routine. Production support documentation from *Chapter 6* is another success definition your team can rely upon.

Along with setting clear goals, accountability relies on providing clearly defined roles on the team.

Establishing roles and responsibilities

It is essential to accountability that each team member have an understanding of what they are and are not responsible for. We detailed the benefits of clear definitions of roles on projects and cross-functional teams in *Chapters 5 and 7*, respectively. In addition to improving product outcomes and reducing conflict, clear roles support an accountable team environment. Defined roles remove ambiguity from determining who is responsible for which aspects of a team's joint efforts.

Without defined roles, teams may be tempted to avoid responsibility for poor outcomes. They can more easily claim confusion or the belief that a missed deliverable was not their job. If everything is everyone's responsibility, it may mean that no one is truly accountable. Consider these potential ambiguities and enact a set of role definitions that allows your team members to understand and live up to the expectations of the team.

If you have tasks that are truly shared and you are not sure how to allocate them, you can have an open discussion with team members where you give them an opportunity to help define their own lines of ownership.

In addition to guidance on what is expected of your team, they need sufficient tools and contextual information to feel confident in owning their objectives.

Providing resources and access to information

The last way to provide guidance for accountability is to ensure your team has the resources and information they need to complete their work. Whether it is problem-framing information for a new project, user research data, metrics on production system health, context on a business decision, or access to screenshare tooling for remote collaboration, teams need to be well equipped to do their best work. Teams are more willing to be accountable when they believe that they have the resources and information they need to do their work well. Engineering managers empower their teams by making sure that is the case.

Take a moment to ask yourself whether you are providing your team with everything they need to take ownership of their work and its outcomes. Are there ways in which you might be a bottleneck for your team? Is there an opportunity to empower them with better tooling, better data, or more transparent communication?

Listen to feedback from your team on what they need from you to do their best work. Empathize with their challenges and work with them to determine how you can best support their efforts.

Providing guidance gives your team the foundation they need to embrace an accountable work environment. With this, your next step is to reinforce accountable behaviors.

Promoting ideal behaviors

Accountable behaviors take root in teams when they are recognized and reinforced. Once you have provided your team with what they need to be accountable, you can shift your focus to supporting and encouraging the right actions. Reinforcing accountable behaviors on your team establishes them as a part of your expectations, culture, and team climate. Frequent reinforcement helps the behaviors become core to how your team operates. Promote accountable behaviors in your day-to-day work by demonstrating those behaviors, coaching, and deliberately sharing authority.

Demonstrating accountability

As with most leadership behaviors, it is best to personally embody the accountability you want to see on your team. Be an example of the highest ideals you want your team to uphold. Be unfailing and unflinching in your personal accountability. And when you fail to be accountable in a situation, as soon as you realize it, step up and be accountable for your failure to be accountable!

To demonstrate accountability, address what behavior was expected, what actually happened, give an apology if appropriate, and specify how you intend to rectify the situation. Use the following guidelines:

- Rather than minimizing leadership shortcomings, take full responsibility for them, such as by saying, "*I know I said we would have this information by this date, but it looks like I was a little overconfident about that. I apologize for the delay and will get back to you with a new timeline or update by the end of this week.*"

- Directly address your own behavioral issues openly and honestly, such as by saying, "*I want to apologize for interrupting the meeting and being called outside. My expectation is that we respect each other's time and every one of us gives our undivided attention during meetings, so I have asked that person not to interrupt in the future.*"

- Take responsibility not just for direct errors, but also errors of omission, ignorance, and negligence, for example, by saying, "*I understand that there was a misunderstanding with this project from multiple people, but I'd like to apologize for my part since I could have made my expectations clearer in the first place. I have revised the project vision and updated our project scoping template with some additional questions that should help us avoid this in the future.*"

- Admit when you've made a mistake due to emotions, such as by saying, "*I've realized that I became too attached to this part of the project and needed to take a step back. I allowed my attachment to the project to overshadow the valid discussion of trade-offs. I'd be glad to revisit the discussion and look at all our options.*"

Having the bravery and fortitude to demonstrate accountability is not always easy, but it is well worth the practice and investment. Along with demonstrating, you can coach your team on the right skills and behaviors.

Coaching for accountability

Use coaching to guide or correct your team's accountability behaviors. Coaching helps to train your team for the right level of nuance and tact in their communications. It is one thing to raise issues and another to do so in a way that is constructive and professional. Coaching can help your team develop the emotional intelligence to give feedback to their peers in a way that is kind and productive.

Let's revisit our coaching steps from *Chapter 9* and apply them to the context of accountability behaviors. The primary accountability coaching scenarios are as follows:

- Not enough personal accountability

- Not enough peer accountability

- Too much accountability

If your aim is to encourage more personal accountability, that is a good case for following the standard coaching approach step by step, as follows:

1. Use their role and established workplace expectations as the basis for a conversation
2. Ask questions and identify needs or gaps
3. Give feedback using radical candor and ask for confirmation of the desired behaviors

For example, if you have an engineer that is shying away from taking responsibility for a mistake they made, dig into why by saying, "*We all take responsibility for our work here, so I'd like to hear your point of view on what happened and where it may have gone wrong. I want to make sure I'm understanding.*" If they shift responsibility, you can continue, "*It sounds to me like there was a miss in following the acceptance criteria here. Am I misunderstanding or what happened with that?*" And if no reasonable justification is given, you provide clear corrective feedback, "*Okay, well we all make mistakes so that is certainly understandable. What I would have loved to see was you recognizing that immediately, taking responsibility, and sharing with the team how you intend to fix the oversight. Can you do that in the future?*"

If your aim is to encourage more peer accountability, you can focus more on the asking and listening step of coaching. Leading with questions helps the coachee to arrive at conclusions themselves and ask themselves questions they may not have considered. For example, if a member of your team is reluctant to challenge their peers directly, you might say, "*Did you ask them about it? What did they say? Did you agree with their answer? That answer doesn't really make sense to me either. Did you ask them why they thought that? What do you think they should have done? Did you ask them why they didn't do that? I think you should ask them that.*" This direct guidance helps your team understand your expectations. Instead of stepping in and resolving the situation yourself, you allow your team to resolve it at the peer level.

On the other hand, you might have engineers who are a little too enthusiastic, intense, or aggressive when holding their peers accountable. This is another case where using more guiding questions is helpful in coaching. For example, a member of your team might be challenging their peers too harshly, so you might say, "*I overheard your feedback to Alex. How was that received? Not well? Why do you think that is? Your goal is to help them improve, right? Well, do you think there might have been a way you could have expressed that where it may have been better received? Yes, I think that is a much better approach.*" Using questions to remind them of their goals helps to encourage rational consideration and empathy during frustrating moments.

Direct guidance through coaching is a powerful means of promoting accountability on your team. Hand in hand with this is creating an atmosphere of shared authority where everyone feels it is appropriate to hold each other accountable.

Sharing authority

To share authority as a leader is to consistently behave in a manner that invites equitable and reciprocal accountability. In other words, it requires adopting a participative leadership style and being open to criticism. Don't just tell your team you want to hold each other accountable, give them the space and freedom to do so. Here are some practices for sharing authority:

- Use collaborative decision-making processes when possible rather than dictating what the team will do

- Use empowering questions to reinforce peer authority on the team; instead of, "*I don't think we should _____,*" say, "*Do you think we should? I agree*"

- Use inclusive and equitable language whenever possible; use more statements with *we* and *us*, reserving *I* for opinion statements rather than decisions

- Welcome and praise peer accountability in group settings; "*I think Alex raises a great point for us to consider, thanks Alex*"

- Own fair criticism that comes your way from the team; if you disagree with the criticism, explain why rationally instead of rebuking or responding with emotion

- Delegate decisions completely to the team when it makes sense to do so

Establishing shared authority with your team gives your engineers the sense that they have agency over their work. They feel empowered to make decisions and resolve team problems themselves.

Along with promoting ideal behaviors in your day-to-day work interactions, you can strengthen accountability by practicing it in key moments of high visibility. For this, we will move on to the next section on how to embrace accountability during challenges.

Accountability in practice

The demands of accountability tend to ebb and flow on teams. In other words, it is more difficult to practice accountability in some situations than others. We all face big accountability moments where we either live up to the ideals of accountable behavior or shy away from the pressure. Highly accountable engineering managers handle challenging situations appropriately, sincerely, and without ego.

Since accountability involves accepting responsibility and owning up to mistakes, it taps into some of our deepest emotions. At times it can be difficult to admit when something is our fault or even partially our fault. We may struggle to come to terms with the ways in which we have allowed or contributed to a failure in our work. It hurts our pride and evokes our fears. We can become so wrapped up in the consequences of a mistake or failure that we become unwilling to face it. So, the first step in practicing accountability is adopting a mindset that doesn't fear failure or shirk blame but owns them as necessary parts of learning and growth.

Further, in most of the difficult situations you will face, you are best served by assuming full responsibility. Provided that you agree that a mistake or failure has occurred that originated with your team, taking an accountable position is more productive than shifting blame to others. Being accountable is perceived well and builds integrity, trust, and authority. Despite mistakes made, accountable people are seen as fair and honest. Shifting blame or otherwise deflecting responsibility creates a weaker impression and demonstrates self-interest over group needs. To fully understand this, let's go through the main scenarios you will face.

Handling the big mistakes

The most difficult times to be accountable are when you have personally made a highly visible mistake, error in judgment, or failed to fulfill a major commitment. Perfection is impossible, so every once in a while, you will be faced with your own mistake at a challenging moment. These are the moments that test our commitment to accountability and demonstrate it to our teams. Because of the attention and reverberating effects of our actions in highly visible situations, these are the most important moments to reinforce accountability and personal integrity.

When you find yourself facing these moments, take a second to prepare yourself. Think through and plan your response. Determine the immediate remediation that you will communicate. Your response could include the same information we introduced in the *Demonstrating accountability* section, but the difference is that it will be under much greater scrutiny, so you'll need to take extra care that your wording and actions are precisely what you intend. Depending on the urgency of the situation, you may want to address it immediately and follow up later with more information.

There may be certain customer-facing situations where you are not permitted to take responsibility for an occurrence because of a lack of absolute certainty and/or liability. Follow guidelines where you must, but know that customers benefit from all of the same advantages of high accountability. Your customers will have greater confidence and trust in you when you are able to own your mistakes.

Handling mistakes made by your engineers

Members of your engineering team may also make highly visible mistakes. When this happens, you would like to see them hold themselves accountable within the team and address it like any other situation. Outside of the team, you will determine how to handle it based on your judgment.

When communicating with higher-ups or those outside of the team, it is generally best for you to take responsibility for errors made by your engineers. You are accountable for their work, for implementing sufficient controls in that work, and for training them well enough to complete their tasks. Rather than singling out a member of your team, you can convey that you did not have sufficient safeguards in place and you are working to fix that immediately. This creates a better impression of your authority and character to your manager, stakeholders, and engineering team. In most high-visibility situations, it is your responsibility to ensure that sufficient remediation is carried out and communicated. You can do this in a manner that encourages team accountability by using the same asking-and-listening method to lead a participative discussion of what happened, how you might have handled it differently, how you can remediate, and how you might avoid it in the future.

On the other hand, there may be times, on occasion, when you determine that it is necessary to allow an engineer to shoulder full accountability for a highly visible problem. You might decide this if an engineer has repeatedly made the same mistake or has acted outside of their duties in an egregious way. It is good to make every attempt to train and handle performance issues within the team, but there may be times when behavior warrants being singled out. Ultimately, it is your judgment call what the right course of action is.

Handling successes

On the other hand, your team will also have some highly visible successes. There will be moments when your team is brought into focus in your organization for a great achievement, response, or performance. These are validating moments that reinforce good practices on your team. Although your entire organization will be listening to what you have to say, in these moments your behavior is most closely watched by your own team.

Demonstrate your belief in the team and shared authority by focusing on them in your response to the situation. Give your team full credit for the success. If certain people were responsible, single them out and congratulate them publicly. Allow them to bask in the moment and feel a sense of accomplishment in their hard work.

It is a good practice to give full credit for success to your team because it builds trust with your team members. It gives them the sense that you are their supporter and fully invested in them and their individual success. It inspires loyalty and confidence that you are not self-serving and that being a member of your team is rewarding. Take advantage of the opportunity to show your team that you are their biggest supporter and promoter.

Ultimately, as an engineering manager, you will get credit for the success no matter what, so focusing on elevating your team in the process is a win-win for you.

Following these practices shows your team and your organization that you take accountability seriously. Although it may be difficult at times, the investment is worthwhile for the whole host of benefits you will receive as you strengthen performance, personal integrity, trust, confidence, psychological safety, and other positive states.

Summary

Accountability is the willingness to take responsibility for your actions and their outcomes. In this chapter, you learned how to build accountability into your team expectations and practices.

The main points of this chapter include the following:

- Signs of low-accountability teams are shifting blame, avoiding addressing issues, escalating all issues to the manager, and apathy

- Signs of high-accountability teams are taking responsibility for actions, holding peers accountable, and resolving disputes within the team

- Teams with high peer accountability are able to move much faster since they are able to resolve their own problems as they arise

- The twin goals of an accountable team culture are to instill beliefs that team members possess **ownership** and **agency** over their work

- Build an accountable team climate by following the *3 Ps*:

  - **Provide**: Ensure team members have guidance and resources to successfully carry out work

  - **Promote**: Recognize and reinforce accountable behaviors on your team

  - **Practice**: Assume accountability during highly visible moments

- Provide guidance by encouraging a shared understanding of work, maintaining clear goals, establishing clear roles, and ensuring access to tools/information

- Promote ideal behaviors by demonstrating personal accountability, coaching for accountability, and sharing authority with a participative leadership style:

  - When you demonstrate accountability, make sure to communicate what behavior was expected, what actually happened, an apology if appropriate, and how specifically you intend to rectify the situation

- Practice during crises by taking extra care to demonstrate full accountability for your own mistakes and assuming responsibility for your team's mistakes

Accountable teams produce better performance, learn faster, and free up managers to do more strategic thinking. Now that you understand how to develop an accountable team climate, we will look at how to safeguard against poor outcomes by managing risk.

Further reading

- *The best teams hold themselves accountable*, hbr.org, 2014. (`https://hbr.org/2014/05/the-best-teams-hold-themselves-accountable`)

- *Does your team have an accountability problem?*, hbr.org, 2020. (`https://hbr.org/2020/02/does-your-team-have-an-accountability-problem`)

- *Accountability in software development*, 2021. (`https://medium.com/@kentbeck_7670/accountability-in-software-development-375d42932813`)

11
Managing Risk

In striving to produce great outcomes, engineering managers must work to avoid disaster scenarios that could arise for their businesses or teams. The practice of avoiding or minimizing disaster scenarios is called *managing risks*.

Risks are any factors that have the potential to lead to failures. In other words, risks are threats to projects and products. For example, a common risk for software development teams is having a lack of redundancy in staffing or in systems that makes the team vulnerable to sudden loss of knowledge or data. Risks may not always materialize into real problems, but to be prepared for when they do, it is good to develop an awareness of risks. The term *managing risk* describes both identifying and responding to risk.

Risk management is its own field of study, so there is a broad body of accumulated knowledge, books, research, and methodology in existence. For our purposes, we will cover basic concepts and ideas applicable to engineering managers' day-to-day work.

In this chapter, we will start with why it is important for engineering managers to know how to manage risk. You will learn how to conquer threats systematically by identifying, prioritizing, communicating, and responding to risk. You will learn when and where it is appropriate to manage risks. By the end of this chapter, you will have the tools to understand your risk landscape and protect your team from looming danger. At the end of this chapter, you will find further resources for growing your understanding of risk management.

This chapter is structured into the following sections:

- Why should you manage risks?
- How do you manage risks?
- When and where should you manage risks?

Let's begin with learning why risk management is an important skill for engineering managers.

Why should you manage risks?

Every so often, we see a story in the news about an organization facing a security exploit because of lax security practices. Some businesses may respond, *We don't see a problem with our practices because we've always done it that way and it has never caused a problem before!* This somewhat baffling mindset is a failure of risk management. It is common for inexperienced or understaffed teams to ignore risk and take the shortest path to project delivery, but as products and teams scale, the dangers of this loom greater. Strong engineering managers must manage both up and down to raise risk awareness and avoid potential disasters.

Depending on your workplace norms, managing risk is often left up to individual managers and teams. When managers fail to detect and address risks, those managers may be accountable if those risks lead to problems or disastrous outcomes. Depending on the severity of the situation, unaddressed risks can lead to substantial failures that become serious career problems for managers, including the possibility of termination. Managing risk helps to avoid common disaster scenarios, such as the following:

- Your systems are compromised by malicious attackers
- You release a catastrophic bug into production
- You fail to deliver a project as promised
- You lose data and cannot recover it
- You lose critical institutional knowledge

Each of these scenarios involves awareness of risks and preparedness for unexpected events. You can circumvent or reduce the likelihood of these situations by implementing a plan to help you recognize and respond to them.

At this stage, you may be thinking, *Isn't this part of project delivery? Didn't we learn about managing risk in Chapter 5?* While project delivery is one application of risk management, it is a broader skill that applies to everything you do as an engineering manager. We will build on what we learned in *Chapter 5* to develop a deeper set of skills and tools for identifying and managing risk, as well as applying techniques beyond the scope of project delivery.

Now that you understand *why* managing risk matters, let's look at *how* we can do so.

How do you manage risks?

Managing risk is a continuous process. Once you develop an awareness of risk, it is something that you will manage every day. Since risks can be so impactful to businesses and teams, it is good to take a rigorous approach to managing them. Take control of risks by identifying them, prioritizing, communicating, and responding.

Identifying risks

Identifying risks, or risk recognition, is the most important step of risk management. Even if you were to stop here, the act of recognizing a threat removes the surprise from it and allows you to have it in mind as you go about your work.

Risks that you identify may not always make sense to further manage. Some risks, such as natural disasters or global events, you may take note of as threats but are not necessarily worth specific mitigation. At this stage, you focus on recognizing and cataloging only, starting with common risks in software engineering.

Examples of common risks

In *Chapter 5*, we learned how to use assumptions and dependencies to identify risks. Let's build on that by revisiting our preceding list of common disasters to see how we might detect risk precursors. For each potential disaster, work backward, asking yourself, *What could cause this to occur?*

- Your systems are compromised by malicious attackers—risks may include the following:

  - Lacking mechanisms to identify vulnerabilities, such as tests and security audits

  - Running out-of-date versions of software

  - Lacking security practices in your team training/culture

  - Lacking knowledge of threat surface

- You release a catastrophic bug into production—risks may include the following:

  - Lacking mechanisms to identify bugs

  - Lacking comprehensive tests for critical systems responsibilities, such as financial calculations

  - Lacking sufficient code review practices

  - Lacking early warning mechanisms, such as canary releases

  - Lacking sufficient real-time system or user metrics

- You fail to deliver a project on the promised timeline—risks may include the following:

  - Lacking staffing redundancies; one person out on sick leave leads to failure

  - Lacking a comprehensive schedule plan; not accounting for holidays

  - Being overly reliant on a particular solution with no backup plan

  - Having **single points of failure** (**SPOFs**) on your team; only one person can accomplish a particular task

- You lose data and cannot recover it—risks may include the following:

  - Lacking sufficient backup and recovery practices

  - Being overly reliant on particular software vendors

- You lose critical institutional knowledge—risks may include the following:

  - Lacking knowledge redundancies

  - Failing to document/codify critical information

  - Documentation has gaps or is not up to date

Similar to the pre-mortem exercise you learned in *Chapter 5*, working backward from an undesirable outcome is a good way to identify risks to look out for. Another way is to consider the different sources of risk.

Sources of risks

Part of risk identification is maintaining knowledge of the major areas where risks may be hiding. Understanding the sources of risk gives you the perspective to see where your team or code base may be vulnerable. Sources of risk specify the environment in which those risks originate, including *physical, social, political, operational, economic, legal,* and *cognitive* (DOI: 10.1108/09566160210431088). The relative weight and management needs of each of these types of risk will be determined by your work context and expectations. Let's have a more detailed look at the different sources of risk:

- **Physical risks** originate in the physical world, such as natural disasters or climate, which may impact engineering teams—for example, through issues with connectivity or data transit.

- **Social risks** concern relationships, behavior, and social structures. They include team emergent states, values, and culture, as well as broad movements such as social unrest and union strikes. These should be assessed continuously since they are always evolving.

- **Political risks** originate with governments and are especially relevant to engineering teams that operate internationally and need to be aware of different national systems, attitudes, and policies.

- **Operational risks** involve the processes and practices in place in your organization, including your team structures, responsibilities, and all aspects of *how* you carry out your work. Consider what risks your processes and practices might introduce.

- **Economic risks** originate with global and local markets, such as economic recessions, trends, and credit policy; these risks can affect an engineering team's budget, staffing, or product outcomes. Consider current economic risks before committing your team to big projects.

- **Legal risks** originate with jurisdictional laws and agreements, such as data privacy and equal access for users with disabilities. Engineering teams need to consider legal risks when determining user interface designs, asset copyright, user data collection, user data storage, and contracts for software and vendors.

- **Cognitive risks** involve the difference between beliefs and reality, such as errors in judgment and cognitive biases. Consider how individual judgment might introduce risk to your work.

Each of these sources brings its own set of potential threats to development teams and their products. Depending on your work context, you may find the majority of your risks in social, operational, and legal sources, but it can be worth considering the full list for good measure. Use these sources as a checklist to remind yourself of the areas to take into account. Now that you understand all the potential sources of risk, you can move on to further practices that help to identify hidden threats.

Further methods for identifying risks

Identifying risks is a mindset that engineering managers develop over time through the practice of thinking about work outcomes from all angles. Here are a few specific exercises that can help bring to light hidden risks.

Trust but verify is an adage that is particularly useful in risk management. This approach helps to identify hidden cognitive risks by confirming that expectations meet reality. Put this saying into practice by monitoring your systems and team performance, as described in *Chapters 6* and *9*. Establish automated checks and perform manual spot checks, especially on new or high-priority work. When discussing these checks with your team, remember to assume the best intentions.

The **pre-mortem exercise** introduced in *Chapter 5* is a powerful exercise for identifying hidden risks for your project and also at the product or system level. Starting from a hypothetical assumption of failure as a brainstorming prompt can help your team to come up with additional risk factors.

Chaos engineering involves deliberately causing controlled failures in order to test how a system responds and recovers. It can be used as a way of identifying risks to systems reliability and to consider areas of focus for improving confidence. There are many different tools available for implementing chaos engineering, such as Gremlin, Chaos Monkey, and **AWS Fault Injection Simulator** (AWS FIS).

Now that you have identified your risks, you can determine how to act on them using prioritization.

Prioritizing risks

Prioritizing risk, or risk rating, is the act of approximating how much time and effort is reasonable to avoid potential risk. Just because you have identified that a risk is present doesn't necessarily mean it is worth your time and effort to remediate it. The main factors that should inform your risk prioritization are *risk tolerance, severity, likelihood*, and *effort*. With these four measures, you can assemble models to guide decisions on the risks you have identified, such as a risk matrix.

Risk tolerance

Risk tolerance, or sometimes risk appetite, is how accepting your organization is of the possibility of failure in pursuit of value. Depending on your workplace goals, growth stage, industry, and customers, accepting some risks may be understood in the pursuit of big rewards. On the other hand, the opposite may be true, such as in financial services, government, client services, or security, where any risks can be a big problem. Work with your leadership team continuously to determine what types of risks are acceptable in your work, if any.

Risk severity

Risk severity is the magnitude of impact of a risk, or how damaging it would be. A common approach is to equate levels of risk severity with particular ranges of bottom-line monetary impact. So, for example, you might determine that you have five levels of severity, ranging from the following:

- **Level 1**: <$1000
- **Level 2**: $1001–10,000
- **Level 3**: $10,001–100,000
- **Level 4**: $100,001–1,000,000
- **Level 5**: >$1,000,001

The figures can indicate how expensive a scenario would be to resolve or how much revenue will be lost if a risk materializes. To calculate ranges, you will need some basic business figures to work from, such as sales or advertising revenues over a period of time. You may or may not be responsible for determining risk severity levels for your team, but if you are, you can work with your leadership to develop an understanding of these monetary impacts and what may be reasonable thresholds.

Alternatively, you can base severity levels on fully qualitative buckets: minimal, low, medium, high, and maximum. Depending on your organizational goals, severity can indicate the number of users impacted or reputational damage rather than the direct cost.

Risk likelihood

Risk likelihood is a measure of how statistically probable or likely to occur a risk is. Risk likelihood is often an estimated qualitative measure, but if you have more specific historical data to work from, that is even better. For example, you might have the organizational knowledge from tracking previous projects that 40% of your new builds end up launching behind schedule. You could then be confident that you have a 40% chance of being behind schedule based on past performance. Use real data probabilities whenever you can source them and make estimations when you cannot.

Risk likelihood estimation can be set to a scale approximating how likely a risk is, such as the following:

- **Rare**: ~2%
- **Unlikely**: ~3–15%

- **Possible**: ~15–50%
- **Likely**: ~51–89%
- **Almost certain**: ~90%

The probability numbers in this example can be adjusted based on the use case and preference. Likelihood levels are often rough estimates because their value is not exact, but is relevant for the purpose of comparison and ranking. It is less important that the likelihood is exactly correct and more important that you establish risk groups for further consideration.

Risk mitigation effort

Risk mitigation effort is a measure of the resources required to sufficiently remediate a risk. General qualitative levels such as the following may be used since mitigation effort is another case where accuracy is less important than comparison and ranking:

- Negligible
- Low
- Medium
- High
- Maximum

It can be helpful to assign ranges to the levels for consistency in usage, such as *Negligible = <1 day's work, Low = 1–2 day's work*, and so on. Alternatively, it may make more sense for your use case to use monetary ranges of cost or to incorporate both of these measures, such as *Negligible = <1 day's work or a cost below $100*.

It can sometimes be a challenge to come up with these estimates without further research, so keep in mind that the levels are broad for a reason, and your aim is to group together like effort with like. For uncertain risks, you can round up your estimate to be cautious.

With that, you now have the information you need to prioritize your risks and complete your assessment.

Completing the risk assessment

In risk management, **risk assessment** refers to the process of completing the first two steps, identifying and prioritizing risks so that they may be acted upon. The method you choose for generating the assessment should be determined by the extent of analysis required by your risk tolerance and business context. However, it is important to note that this process, much as with project planning, can never be entirely foolproof. Therefore, it is advisable to timebox it, ensuring it remains within an acceptable timeframe. If you are in a risk-tolerant environment or have limited time, you may prefer a simple assessment, while those in environments that must be careful and exacting about risks and have time to accommodate can use deeper analysis methods. Consider what may be the best fit for your organization and confirm that with your leadership.

The simplest approach to risk assessment is to focus mainly on the severity and likelihood of the risks to produce a high/low quadrant, as follows:

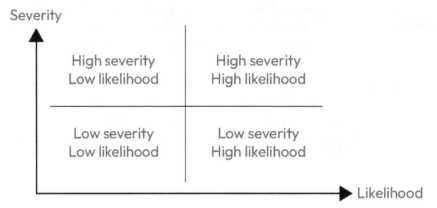

Figure 11.1 – Risk assessment with severity and likelihood

With this approach, the assessment goal is to give top priority to the high-severity, high-likelihood risks, lowest priority to the low-severity, low-likelihood risks, and moderate priority to everything else. Using the previous example scales given, that could look something like this:

Risk assessment and mitigation plan for web team:

- *Mitigate all high-priority risks (sev. level: 3+ and likelihood: likely+)*
- *Monitor all low-priority risks (sev. level: <3 and likelihood: <likely)*
- *Mitigate moderate priority risks with low or negligible effort*
- *Review remaining moderate priority risks with leadership for further direction*

This approach allows you to bucket risks and make a clear plan for what you will mitigate and why. Avoid the *McNamara fallacy* (see *Chapter 9*) by double-checking your assessment to determine if there are risks that make sense to mitigate despite not meeting the formula criteria.

The high/low quadrant is the most basic form of a risk matrix. Deeper methods of risk assessment involve using more detailed risk matrix formulas, additional models, and weighting individual factors differently. Detailed risk matrices take the same approach as previously, but they introduce more boxes and gradations for further classification and insight. Here are some deeper methods to consider when desired:

- Full risk matrices have pros and cons well described by Julian Talbot (https://www.juliantalbot.com/post/2018/07/31/whats-right-with-risk-matrices)

- The **Open Worldwide Application Security Project** (**OWASP**) community risk rating methodology is another matrix approach (`https://owasp.org/www-community/OWASP_Risk_Rating_Methodology`)

- The bow-tie method can be used to model particular disaster scenarios in a visual format (`https://www.juliantalbot.com/post/risk-bow-tie-method`)

You now have a complete risk assessment to guide team discussions, decisions, and mitigation efforts. From here, you are ready to report risks outward.

Communicating and responding to risks

In *Chapter 5*, we briefly introduced risk communication and remediation, with a focus on timeliness and audience appropriateness. Beyond urgency and audience framing, there is more you can do to add context and clarity to your risk reporting. Depending on your role in managing risks for your team, this may be as simple as handing your recommendations off to supporting functions in your organization. For those engineering managers who need to progress the risk management further themselves, here are a few tools to aid in communicating and remediating risks.

Use the 4 Ts to plan a risk response

When planning a response or mitigation for risk, consider a variety of options by using the 4 Ts: *tolerate, treat, transfer, terminate*, elaborated on as follows (from *Fundamentals of Risk Management, Paul Hopkin and Clive Thompson, Kogan Page, 2021* at `https://www.koganpage.com/risk-compliance/fundamentals-of-risk-management-9781398602861`):

- **Tolerate**: Accept the risk; acknowledge it but take no action

- **Treat**: Take action to reduce the impact or exposure

- **Transfer**: Shift the risk to another party

- **Terminate**: Eliminate the source of the risk

Tolerating a risk may be your choice when the likelihood is extremely low and the severity is low as well. Treating a risk may be how you will respond to the majority of risks you identify, by taking action to avoid or reduce the damage of the risk. Transferring the risk can make sense in some cases—for example, avoiding legal responsibility for compliance by engaging a software vendor or third-party service to own that aspect of your product/system, such as a cookie consent pop-up service. Terminating means that you abandon your plans to build the feature that introduces the risk, such as deciding not to collect user data to avoid compliance regulations. This may be your choice if the severity and effort are too high, leading you to reconsider the product approach to choose something with less risk involved.

After selecting your preferred response to risks, validate that the response is sufficient.

Use the 4 As to validate your risk response plan

Once you have drafted a plan of action for risks, it's a good idea to circle back and confirm that you are actually meeting your response target. Validating these plans is useful because it is not uncommon for risk responses to fail to accomplish their intended goals. You can validate your response by using the 4 As: *appropriate*, *agreed*, *actionable*, *achievable*, as follows (you can read more details at `https://www.juliantalbot.com/post/2018/03/06/so-what-whats-so-important-that-its-worth-writing-a-whole-book-about`):

- **Appropriate**: Is the response reasonable for the level of impact the risk may have? Does it pass the common sense test?

- **Agreed**: Do your team, cross-functional partners, stakeholders, and/or leadership agree with the response?

- **Actionable**: Is it clear how the response will be carried out?

- **Achievable**: Is it a response that can be realistically completed with the resources you have?

Validating your risk response provides a *sanity check* to reaffirm your actions are lining up with your intentions. With that, you have completed all the steps and have the tools for managing risk. All that is left is to ask yourself, should you be managing risk?

When and where should you manage risk?

Given that risk management is its own field independent of engineering management, it is not always appropriate for engineering managers to take on the responsibility. Risk management can be a full-time job on its own in many cases. You will need to determine the right level of risk management to engage in. To avoid setting yourself up for failure, don't take full responsibility for risk management when it means you will have to sacrifice more essential duties as an engineering manager.

Some level of risk management can always be assumed as a leader within an organization. Understanding and managing the common risks in software is a good baseline to have in hand as an engineering manager.

In lean workplaces, such as early-stage start-ups with few employees and support roles, you may consider owning project risk management. If you do so, take care to balance it with your other responsibilities by choosing simple assessment methods and effort-preserving risk responses, such as terminating risks from unnecessary or questionable-value product decisions. You might also be able to delegate risk management to a member of your team to distribute the workload.

In some situations, you may need to abstain from risk management in service to your primary role as an engineering manager. This is sometimes known as *allowing things to fail*. While you never *want* things to fail, there are times when failure is necessary to call attention to plan deficiencies, unrealistic expectations, or a lack of necessary support. For example, if your team is understaffed for its objectives and you have communicated this clearly to stakeholders, you might consider working 80-hour weeks to mitigate the timeline risk on your projects, but you would have to ask yourself how sustainable that is and what sort of precedent it would set. It might be a better choice to tolerate the risk and miss the deadline, illustrating that it was not achievable without better staffing. Especially early in your career as an engineering manager, it takes bravery to allow things to fail, but at times it can be a better path to resolution.

With these guidelines in mind, work to develop your own sense of when and where managing risk is appropriate.

Summary

Risk management is a valuable skill and field to be aware of as a leader. Borrowing techniques from risk management can help engineering managers be successful and avoid disastrous outcomes.

Here are some key takeaways from this chapter:

- Managing risk is an approach and practice to avoid potential disaster scenarios.
- Managing risk helps you to assess and communicate what disaster scenarios may arise in your organization and why they should or should not be addressed.
- The steps to managing risks include identifying, prioritizing, communicating, and responding:
  - To identify risks, follow these guidelines:
    - Familiarize yourself with common software engineering risks
    - Look to the various sources of risks: physical, social, political, operational, economic, legal, cognitive
    - Employ automated and manual checks on your team's work
    - Use the pre-mortem brainstorming exercise
    - Implement chaos engineering
  - To prioritize risks, follow these guidelines:
    - Determine the right level of risk tolerance
    - Create measures for severity, likelihood, and mitigation effort of risks
    - Sort risks into priorities or ratings using the measures you created

- To communicate and respond to risks, follow these guidelines:

  - Convey risks with timeliness and audience focus

  - Consider whether to tolerate, treat, transfer, or terminate the risk

  - Validate that responses are appropriate, agreed, actionable, and achievable

- Decide when and where it is appropriate to manage risks based on organizational factors. Always manage for common software engineering risks such as security and catastrophic software bugs.

Now that you wield the knowledge and tools to manage risks, that concludes *Part 3* of this book. In *Part 4*, you will learn how to lead during times of change, beginning with resilient leadership.

Further reading

- *Hopkin, Paul and Clive Thompson. Fundamentals of Risk Management: Understanding, Evaluating and Implementing Effective Risk Management. Kogan Page Publishers, 2021.* (`https://www.koganpage.com/risk-compliance/fundamentals-of-risk-management-9781398602861`)

- *Weick, Karl E. and Kathleen M. Sutcliffe. Managing the Unexpected: Sustained Performance in a Complex World. John Wiley & Sons, 2015.* (`https://www.wiley.com/en-us/Managing+the+Unexpected%3A+Sustained+Performance+in+a+Complex+World%2C+3rd+Edition-p-9781118862414`)

- An explanation of canary releases by Martin Fowler (`https://martinfowler.com/bliki/CanaryRelease.html`)

- *DeMarco, Tom, and Tim Lister. Waltzing with Bears: Managing Risk on Software Projects. Addison-Wesley, 2013.* (`http://www.dorsethouse.com/books/waltz.html`)

Part 4: Transitioning

In this part, you will learn how engineering managers lead the transitions that occur. These chapters focus on preparing for and navigating changes to your team. You will learn how to cultivate resilience, recruit and hire for your team, and lead your team through organizational changes.

This part has the following chapters:

- *Chapter 12, Resilient Leadership*
- *Chapter 13, Scaling Your Team*
- *Chapter 14, Changing Priorities, Company Pivots, and Reorgs*

12

Resilient Leadership

The pace of growth and change in modern business is staggering. This pace makes the work of engineering managers exciting, with new challenges and solutions every day. At times, the constant change can become a burden, disruptive to ourselves and our engineering teams. We may ask ourselves, what can we do as engineering managers to produce good outcomes for our teams when it feels like the sands are always shifting beneath our feet?

In *Part 4* of this book, we will dive into the role of managing change within software teams and businesses, starting with how to build resilient teams. In your career, you have probably seen colleagues react to change in all sorts of ways. Change can have destructive effects, or it can allow us to reinvent and revitalize. As engineering managers, we can better support our teams by ensuring that they can adapt when changes come along.

In this chapter, you will learn how to prepare teams for inevitable change by making them—and yourself—more resilient. You will learn why resilience matters and why teams with it are more successful. You will learn how to build a resilient culture for your team so that its members are more confident when facing the unknown. You will also learn how to introduce change and transition your teams in a way that is easier for them to accept. By the end of this chapter, you will understand why resilience is a crucial team characteristic and how to cultivate it in your team.

This chapter is organized into the following high-level resilient leadership topics:

- Introducing resilient teams
- Preparing your team for change
- Preparing change for your team

Let's start by learning why we need resilience in the first place.

Introducing resilient teams

Software development teams face change every day. These changes range from the micro level, such as adjustments to project requirements, to the macro level, such as changes in company direction or goals. When teams overreact to change with emotion by losing confidence, motivation, and momentum, those teams are described as **fragile**. When teams are more adaptable and react to change with curiosity and open-mindedness, they are **resilient** teams. The level of change a team can adapt to indicates how fragile or resilient that team is.

The most fragile teams are easily disrupted by relatively trivial changes. They have a limited comfort zone in which they are capable of working. They cling to the familiar and react to change with fear, annoyance, and other strong emotions. They may have an overly rigid view of their workplace and role.

The most resilient teams can consider and navigate significant changes. They can focus on the big picture and take into account how change may be beneficial in the long run. They react to change by seeking to understand why it is happening, what it will mean for them, and how they may contribute.

Why do resilient teams matter?

Resilient teams fare better because changes are inevitable and emotional responses to them can be unpredictable. Over days, months, and years, you can guarantee that your team will face new projects, new directions, new methods, new areas of focus, and new leaders with new ideas. When teams are too attached to certain ways of doing things, change has the power to shake your team to its very core.

Change taps into deep emotions that you cannot always be aware of as an engineering manager. This can be surprising at times. For example, you might see someone have an utter meltdown over a seat assignment change when their desk is moved to another location in a shared office. You might have had no idea that this person was strongly attached to their desk and considered it an important part of their work benefit. The perception of taking that away from them could be met with high resistance or even a reevaluation of their job satisfaction. And that is just for a seat change. Bigger organizational changes have the potential to evoke huge emotional responses.

Engineering managers can have greater success in their roles by developing skills to help their teams (and themselves) weather these changes.

You might be wondering, why can't we just staff our teams only with those who are comfortable with change? While it's great if you can do so, that can be a hard goal to attain and maintain. Developing the ability to increase the resilience of your teams is something that will serve you well throughout your career at times when simply restaffing may not be possible.

The engineering manager's role in creating resilient teams

In *Chapter 5*, we introduced *change readiness* as an important foundational expectation for teams beginning a project. The more that teams can internalize that change is normal and to be expected, the better they will be prepared to adapt when changes come along. This is true of changes in projects as well as major organizational changes that reshape companies. Engineering managers are in a position to develop change readiness and shape the effects of change on their teams.

To produce the best outcomes, engineering managers can develop the ability to modulate change, guiding the flow of change in both directions. In other words, engineering managers support change readiness by both preparing their teams for change and preparing change for their teams. In the following sections, we will learn how to master the art of change readiness in both directions.

Preparing your team for change

During their work and communications, engineering managers have the opportunity to prepare for future change before it comes along. First, you need to prepare yourself for uncertainty and change. Next, you must work on developing the right team emergent states to form a resilient culture on your team. Finally, you must introduce practices and processes to create resilient habits. Let's go through each of these.

Managing yourself

When you think about preparing your team for future change, it can be easy to forget that you are a part of that team and must prepare yourself. Good engineering managers often think about their team's needs before their own, but since change connects us with deep emotions that are not always obvious, we must take measures to manage and care for ourselves so that we can effectively provide for our teams. To begin, let's review some ways in which we can deliberately increase our resilience.

What is your purpose?

In *Chapter 6* and throughout this book, you have learned how powerful it is to give work meaning and instill a sense of purpose in the work carried out by your team. This is as true for you as it is for the rest of your team. The first step to preparing yourself for future change is to check in and assess your sense of purpose. What are your reasons for being an engineering manager and what drives you to do so within your organization? What do you hope to achieve for yourself and others? Write down the answers to these questions for reflection. Display them prominently in your work area to remind yourself daily why you do what you do and what you are working toward on a personal and professional level.

Without a strong sense of purpose, it becomes harder to detach from your own emotions about the work you do. You will be more susceptible to the **sunk costs fallacy**, believing that something should be continued to justify its own historical cost. Since it is entirely natural to feel a sense of pride and attachment to the things you build and create, it is a good practice to rebalance that attachment by continuously framing your work in the context of the bigger picture. It may be that something you have lovingly worked on for years now best serves you as a set of learnings that you take with you when pivoting to an entirely new approach.

Leverage your purpose to help give you the courage to face and adopt transformative change as it comes along. Focusing on your purpose reduces fatigue and burnout. It also grounds you and provides something to hold on to in times of uncertainty and change. With the big picture consistently in focus, you can remain centered on what you are working toward and be less thrown by changes in how you get there. You will have the right perspective to consider change within the context of high-level goals rather than specific methods. You can more easily consider the benefits of change and how it may advance your purpose. You will be able to assess what questions you have about the changes from a more generative point of view.

Building your support network

The next method of increasing your resiliency is to build your support network. Personal fortitude can get you far, but with support, you can always go further. Give yourself a network of peers and mentors to seek out during times of change and uncertainty. Work on filling the following roles in your support network at a minimum:

- **Your manager**: Developing a strong relationship and rapport with your manager helps give valuable context and guidance during times of change. If possible, create a dialog with them to exchange viewpoints on different changes, transitions, and organizational happenings. Depending on your manager, they may not always be a prominent figure in your support network, but where possible, this is an ideal source of support.

- **An external peer mentor**: As a counterpoint to what you hear and discuss with leaders in your organization, it is valuable to find a mentor from outside your company to discuss ideas and approaches. While you will want to avoid discussing things such as trade secrets and proprietary information, you'll be able to hash out thoughts on managerial processes and transition strategy. Being able to go to someone with zero involvement or investment in the situation in your workplace is an incredibly useful perspective. You can seek out mentors like this via community groups (local meetup circles), previous coworkers, and online interest groups (Reddit, Discord, and so on). Start a conversation by telling them you would love to talk shop sometime.

- **A senior leader:** Senior leaders who are multiple levels above you, whether internal or external, often have very different perspectives and contexts to offer line managers. It can be harder to cultivate mentoring relationships with them, but it is worthwhile to look for opportunities to do so. Oftentimes, directly telling someone that you are working toward developing more leadership skills and would love the opportunity to ask them a few questions can be enough to start a mentoring relationship with a senior leader.

- **A cross-functional mentor:** Similar to external mentors, cross-functional mentors can offer a different mix of skills, ideas, and viewpoints as you work through change scenarios. It can be enlightening to learn where those viewpoints are similar and where they may be completely different. This is another case where you can seek out mentoring relationships from your partner teams by asking directly and seeing where potential mentors have interest and availability to work with you.

- **Trusted peers:** Having both cross-team and cross-functional peers in your support network is also valuable. Within your company, peers from other engineering teams, other functions, or other products may be more comfortable sharing candid perspectives with you and comparing notes. These relationships provide resilience as a source of differing viewpoints and as a peer outlet for mutual expression or venting. One thing to avoid here is spending too much time with internal peers and creating an echo chamber.

Building a network of supportive voices gives you a range of advisors to consult and discuss ideas. Having that level of support is particularly useful when you are working through changes, transitions, and uncertainty that you may not have fully formed your opinion on. It is immensely valuable to have others to bounce ideas off of and share perspectives. Whether they agree or disagree with your assessments of situations, they provide good points and counterpoints to add depth to your thinking.

Practicing self-care

There may be trying times in your work as an engineering manager when you attempt to get through challenges by sheer force of will, but it is hard to be resilient if you are not taking care of your basic human needs. Practicing self-care means setting aside time to be healthy and well adjusted in your life. In other words, self-care involves deliberately maintaining your mind and body so that you can be the best version of yourself. The simplest ways to practice self-care are to make sure you are getting enough rest, exercise, and nutrition. It may sound basic, but it can be easy to ignore these things when we are fixated on particular goals that feel more urgent in the moment.

When we don't take time for self-care, we may initially feel like we are creating more time for work and being more productive, but the lack of restorative time degrades our output. The time we spend working becomes less potent and less effective. Problems are harder to solve because our brain is not operating at its true capacity when we are not well-rested and healthy. Emotions run higher because we are not experiencing the catharsis of exercise. Your best work requires you to be your best self. Like all systems, your body needs regular maintenance and care to perform at its best. This can be most acute during stressful situations where you need resilience.

You owe it to yourself to take equally good care of your mind and body as you do for your work and your team. Make accommodations to get a full night's sleep each night, perform some physical activity, and eat a balanced diet. Take up a hobby that relaxes you after work. Read a book you find stimulating. These small choices make a big difference in your resilience.

While self-care is foundational to day-to-day resiliency, sometimes, you will need a little more to get you through a big change moment.

Hyping up and calming down

Another aspect of managing yourself is the ability to control your energy level. You might be going into a situation where you know you have to make an unpopular decision, deliver difficult news, or face an unyielding critic. During these times, it can be helpful to hype yourself up, taking a moment to strengthen your confidence and energy for the task at hand. Alternatively, you may need to calm yourself down, focus, and destress to clear your head. It might feel a little weird or silly, but hyping up or calming down can help you perform better in clutch moments. Depending on your temperament and the situation, you may be more inclined to need to hype yourself up or calm yourself down during a moment of transition.

Here are some ways you can hype yourself up:

- Listen to a "hype up" song or playlist that makes you feel more confident
- Physical exertion; do a little bit of exercise to get your heart rate up, such as jumping jacks or jogging in a circle
- **Power posing** works for some people—take up space by standing with your arms outreached in a V or take another pose that feels powerful to you
- Repeat a mantra to yourself, such as "I can do this!"

Now, let's look at some ways you can calm yourself down:

- Have a moment of silence, taking a few deep breaths with your eyes closed
- Shaking it out—release bodily tension by standing and letting your arms and shoulders go limp, then shake your shoulders so your arms sway loosely and your body relaxes
- Use a calming mantra, such as "I believe in the work I am doing and I am enough"

Start with these and see what works best for you. It may feel a little strange at first, but channeling your energy through preparation rituals can be a powerful way to overcome anxiety and emotion.

Now that you understand how to manage yourself, you can shift your attention to developing a resilient culture on your team.

Building a resilient culture

In *Chapter 9*, we learned how to bring about positive team emergent states and team climates to reshape the norms of our teams. Resilience is another positive team climate that we can build into our culture and expectations over time. As we learned previously, we can work toward establishing a team climate through specific precursors, along with example setting and behavior reinforcement. But first, let's see what underlying qualities are known to make teams more resilient.

Resilient qualities

In response to rapid change in business, resilience and change preparedness have been increasingly studied, theorized, and written about. While no single view of the underlying precursors is widely agreed upon, there are some consistently observed qualities of resilient teams. Some of these are team emergent states themselves, which we identified earlier as performance building blocks, while others are new to us:

- **Trust:** With its foundational role in vulnerability and risk, it makes sense that trust is also foundational in providing resilience against change and uncertainty

- **Compassion and caring:** Strong relationships within a team bolster confidence during times of change

- **Helping to lift each other up:** The most resilient teams treat resiliency as a group responsibility rather than an individual trait

- **Candor and willingness to give feedback:** Resilient teams have the psychological safety to ask honest questions and give honest opinions

- **Being empowered, resourceful, and action-oriented:** When the team believes they can resolve their problems, change and uncertainty are less daunting

- **Humility:** The willingness to ask for and accept help can be a stumbling point for team resilience, so humility can help prevent this

- **Confidence or belief in the team:** This is another foundational trait since our beliefs often dictate our effort and willingness to go along with change

- **Shared mental models:** These are relevant because change and uncertainty increase the need for team coordination and that coordination is easier when our understanding of work is consistent

- **The ability to improvise:** Change typically cascades through processes and practices, so improvisation during transitions helps teams be more successful

- **Shared values:** Having established values that do not change provides stability and grounding when work methods or goals are changing

Some of these qualities are familiar as team emergent states, which you have an idea of how to work toward (and may have done so already). Other resilient qualities may leave you wondering how to approach them. How do you increase humility? Or improvisation? For that, let's go through a simple approach to encouraging new behavior.

Encouraging behaviors on your team

An acronym to remind you how to encourage any behavior on your team is **EDIT: express, demonstrate, incorporate, and pay tribute**. *Express* the behavior as an explicit expectation. *Demonstrate* or model the behavior with your own words and actions. *Incorporate* the behavior into existing processes for onboarding, project work, meetings, hiring, and so on. Pay *tribute* to or recognize this behavior by praising, calling attention to, or rewarding it when it is demonstrated by members of the team.

For example, let's say you want to work toward building more trust within your team. You would express the importance of trust to the team in writing or vocally. You might address it directly by stating to the team, *I want us to be able to trust each other*, or *I believe it is important that we trust each other to each do our part*. You would demonstrate it with careful attention to being trusting and trustworthy, following through on your commitments and expectations. You could incorporate trust into your team's practices by updating documentation, such as including in your onboarding docs and stating *we trust engineers to do X and to raise issues with Y*. You might pay tribute to instances where team members followed through on their commitments or showed dependable behavior by praising them in a group setting; for example, *I want to thank this person for showing exceptional trustworthiness this week by _____*.

Another way to think about encouraging behaviors is by using James Clear's *Identity-Based Habits* (`https://jamesclear.com/identity-based-habits`). He asserts that the best way to change behavior is by incorporating that behavior into your identity. In other words, instead of emphasizing the actions, emphasize being the type of person that carries out those actions. For example, to affirm that trust is part of the team's identity, you might establish a team value of trustworthiness.

Introducing values

Providing a shared set of values is a common practice across software engineering departments and teams. While your purpose and vision impart motivation and alignment to teams, values add another important piece of the puzzle, giving the team a shared identity. They provide us with a sense of who we are and how we work by answering the question, what characteristics do we value in our team?

Since changes that come along may alter the team's vision and may also affect their purpose, shared values are particularly important to team resilience. Values can provide a sense of permanence, identity, and grounding during uncertain or tumultuous times.

Engineering values can range from technical to purely cultural. They are typically written as statements or imperatives, such as *"fast feedback," "value unique perspectives," "simple is better than clever,"* or *"we are passionate about our work."* Values can be structured as a word, a phrase, a sentence, or a phrase with further explanation. It's good to have at least several values in a set, but not so many that it becomes difficult to remember them all.

Depending on your company's size and stage, you may have well-established engineering values or it may be completely up to you to create them. If you have a broad set of company values, it may still benefit you to have a more succinct set of team values that more specifically address your team's work and challenges. Ask yourself if the current set of values you have encompasses the full set of behaviors and expectations you have for your team.

Authoring your values

Well-written values should be contextual, aspirational, and resonant. They are contextual similarly to other practices we have introduced in this book, meaning they should be respective to situational and organizational needs. They should be aspirational in describing the ideals of the team. They should be resonant and pithy in describing those ideals in a way that is both potent and memorable. Values you establish can have many goals and uses, but to start, we will focus on writing them in the context of growing resilience.

As an exercise, let's say we are writing a set of new engineering values for a startup where the environment is lean and fast-paced and our product focus is on finding product market fit in a competitive landscape. As the engineering manager for this team, we understand that our context requires agility and being prepared for the possibility of big shifts and pivots on the path to growing our user base. Fortunately, we have a team of experienced engineers that have the skills to meet a wide range of product challenges. To build a resilient culture for our team, we can start by looking at the resilient qualities. We could decide, *my team is highly capable with good relationships and trust within the team, but I believe we are sometimes lacking in helping to lift each other up, having humility, and improvisation.* Pause for a moment before continuing and consider what values you might author to reinforce those behaviors.

To encourage our team to view resilience as a group responsibility and lift each other up, we could write a value called *Better together. Better together—we believe we can go further by going together.* To encourage humility in a team with big talent and big egos, we could write a value called *Work with humility. Work with humility—we believe there is always something we can learn from each other when we listen.* To encourage improvisation on our team, we could write a value called *Be creative. Be creative—we believe there is a solution to every problem.* These values could be introduced into a larger complete set.

Once your set of values has been authored and agreed upon, make sure you put them into practice by using the EDIT behavior method. Values are meaningless without following through and upholding them in your day-to-day behaviors and processes.

With this foundation for encouraging the right behaviors, you can further support your team by providing deliberate cues for resilient actions.

Building resilient habits

While you go about the work of encouraging resilient behavior on your team, you can also implement the means to prompt those behaviors directly. When establishing new norms and expectations, it helps to provide clear avenues and situations for them to take root. Build resilience in your team by introducing candor breaks, holding training exercises, and providing focused outlets.

Taking candor breaks

Candor and feedback are qualities central to resilient teams. It is one thing to inform people that they may speak candidly and give feedback, but for some of our colleagues, getting participation in this can be challenging. One solution for this is holding specific sessions for feedback called **candor breaks** (https://hbr.org/2021/01/7-strategies-to-build-a-more-resilient-team).

Candor breaks are a way to pause a discussion and seek more genuine feedback in the moment. Providing a specific venue and prompt for candid thoughts helps encourage and reinforce the expectation, especially among those who may be less likely to share their point of view. The eventual goal is to get to the point where you no longer need candor breaks because the team has the psychological safety to speak candidly in the normal context of discourse.

There are two main approaches to candor breaks. The first is giving a prompt to the whole group, such as *Let's take a candor break; what is not being said here?* You may even ask the question in turn to each group member. The second method is to move into smaller breakout groups to encourage candid discussion among peers—for example, *I think we need to dig deeper into this; let's take a candor break. You and you partner up and exchange your candid thoughts on this topic for 10 minutes and then we will regroup to share what was discussed.* With breakout groups, it is best to partner up those with the same level of seniority to balance the power dynamic.

To introduce candor breaks to your team, make a simple announcement about the intent: *In support of our values, I want to raise the amount of feedback and honest opinions we exchange as a team, so I am introducing a new practice called candor breaks. Any time we discuss anything, you can call a candor break if you think others are holding back their real opinions.*

Candor breaks are a great way to prompt deeper discussions, but resilient teams need more than candor. For that, we can introduce training exercises.

Training exercises

The way to get better at anything is through practice. This includes getting better at change readiness. One of the reasons why change can be scary is because it may feel sudden and unanticipated. We can reduce this feeling by "practicing change" or holding sessions to rehearse and gamify hypothetical changes. This is an occasional practice that can increase resilience by encouraging everyone's thought processes to be as flexible as possible.

To introduce resilience training exercises, start by considering the best way to relate them to your values and/or vision when it comes time to present to the team. With that in mind, write up a series of *what-if* change prompts as your hypothetical scenarios, such as the following:

- What if we needed to accommodate this new product use case?

- What if our product vision changed from _____ to _____?

- What if we could no longer rely on _____ technology in our systems?

- What if we had to complete _____ project in half the time?

Training exercises can run the risk of feeling cumbersome when we have other demands on our time, so it is best to use them sparingly and keep them focused and fun. For example, let's assume you are the manager of a fully remote team of engineers and you want to introduce training exercises to build their resilience. You might introduce them by saying, "*As you know, one of our values as a team is Be Creative. In support of this value, I am setting aside a couple of hours for us to practice creativity by working through hypothetical problems and solutions. On "ready-set-go day," we'll take the afternoon to go through 2 hours of exercises and see who can come up with the most creative solutions to a set of hypothetical product problems. Then, we'll debrief and chat about how we did while enjoying some refreshments I am sending your way in the mail.*"

With a bit of incentivization, training exercises can build the habit of responding to novel change scenarios with resilience and confidence.

Similar to how candor breaks provide the space for feedback, you can build resilient habits by creating other outlets for your team's thoughts and questions.

Creating outlets

When you provide healthy outlets for your team's emotions, you increase resilience. Creating outlets for non-specific self-expression can provide a pressure release valve that helps your team manage their stress or send up a flag when they need help. Providing outlets teaches your team how to channel stress in productive or at least non-destructive ways. Productive outlets can include retrospective meetings, informal chat rooms, and connections with senior leaders.

Retrospective meetings are forums for reflecting, discussing problems, suggesting improvements, and raising issues from recent work. They are commonly held for projects and project sprints but they can also be held in other contexts, such as quarterly or monthly engineering discussions, or in the context of specific organizational changes. Retrospectives provide an outlet for freeform expression but should be geared toward solutions to address what is raised.

Group chat rooms that are available in your team's persistent chat application are excellent outlets for everything from serious questions to general venting. You can be extremely specific when setting these up and directing your team to use them. For example, you might set up a room for #*[team]-questions* that you check periodically and comment on when appropriate. On the other hand, you might set up a room for #*screaming-into-the-void* where engineers can go to "yell" their frustrations during a stressful moment and receive a chatbot response. These small outlets can be excellent coping mechanisms for teams.

Connections with senior leaders are useful as outlets beyond the bounds of your team. Since workplace scenarios will often extend beyond your team, it can be helpful to advocate for ways that engineers can connect with senior leaders in the company. While it may not be in your direct control, it is worth asking of your leadership team. Methods for this can include anonymous question collection (with or without voting), designated chat rooms for #*ask-leadership*, or just Q&A during all-hands company meetings. Partner with your manager and company leadership to advocate for these connections.

Use these outlets to help build healthy habits that channel the team's energy and relieve stress—and come up with some of your own.

With that, you have a solid understanding of how to prepare your team for change and can start learning how to prepare change for your team.

Preparing change for your team

There are right ways and wrong ways to carry out changes in your team. During a change scenario, the leadership you demonstrate can make the change easier or harder for your team members. You might be leading the change yourself, privy to changes in advance of your team, or changes may become known to you and your team simultaneously. In each of these cases, your team will be watching you closely to see how you handle the situation. There will be no second chances for these moments, so be prepared to handle the situation in a way you are satisfied with.

Change communication

Whether the change is completely under your control or outside of your control, being able to communicate it clearly can make a big difference in producing a successful transition. Ideally, you should be able to include each of the following change characteristics in your communication:

- **Vision**: A high-level understanding of what is changing and why
- **Strategy**: The plan for transitioning or implementing the change
- **Motivation**: Finding a way to personalize the benefits of change
- **Flexibility**: Guidance on how firm the change is, including how they may or may not be able to influence it
- **Immediate actions**: What they need to do immediately or how they might contribute to the change effort

Each of these pieces of information plays a role in orienting the team to the change and making the communication more digestible. This helps the team feel that they understand what, why, and how the changes are taking place.

If you find that you are fully on the receiving end of a change announcement and receiving the information for the first time in a group setting with your team, you can phrase these change characteristics as questions to the announcing leaders. For example, if your department leadership announces a new vision and strategy, you might seek this information for your team by asking, "*Is there flexibility in the rollout plan for this change if we need to wrap up our project?*" and "*How can engineering teams contribute and give feedback during the transition?*"

In addition to providing the necessary information in your change communication, there are some change leadership guidelines you can follow to soften the impact of a transition.

Change leadership

As engineering managers, we are often more senior and have had longer careers where we have experienced a range of situations. We have probably been exposed to different types of change and different approaches to our work and problem-solving. Our teams, on the other hand, might have some engineers who are earlier in their careers. When you are early in your career, you have less exposure, so the prospect of change can be much more alarming. It is good to remember that the magnitude of change may feel much more serious to our engineers than it does to ourselves. Be respectful of that by taking extra care when you manage change, including the following:

- Take the change seriously; don't be too casual or flippant about change because that may come across as insincere

- Don't be afraid to adopt a more directive leadership style during times of change, since this is the time when alignment through clear direction is most appreciated (see *Chapter 2*)

- Overcommunicate the change; reinforce what is happening and what is expected of each team member

- Rally your supporters; if some team members are excited to contribute to the change, leverage that to help make the transition successful

- Be receptive by listening to feedback and then doing something about it; if legitimate concerns are raised, be sure to act on them accordingly

- Aim for a quick win; look for ways to reinforce the change by demonstrating benefits early in the process

With these change leadership practices, your change will be in the best position to gain momentum and become standard practice for your team.

Summary

The pace of change in business is rapid. Engineering managers that use resilient leadership to navigate and embrace change scenarios gain a competitive edge for themselves and their teams:

- Engineering teams with resilience react to change with curiosity rather than fear and emotion

- Engineering managers increase resilience in their teams by simultaneously preparing their teams for change and preparing change for their teams

- Prepare your team for change by managing yourself, building a resilient team culture, and building resilient team habits:

  - Manage yourself through affirming your purpose, building a support network, practicing self-care, modulating your energy levels, and avoiding taking on too much due to bias

  - Start building a resilient culture with foundational team characteristics: trust, compassion, the ability to lift each other up, candor, empowerment, humility, confidence, shared mental models, improvisation, and shared values

  - Encourage resilient behavior by expressing explicit expectations, demonstrating it yourself, incorporating it into your practices and processes, and paying tribute to it when the behavior is shown by your team

  - A powerful means of introducing a new set of ideal behaviors is by establishing shared values that encompass the permanent expectations of the team

  - Shared values should be contextual to business needs, aspirational to ideals, and resonant in expressing ideas clearly and memorably

  - Reinforce resilient behavior by integrating resilient practices such as candor breaks, training exercises, and non-specific team outlets

- Prepare change for your team with careful change communication and strong leadership:

  - Change communication should orient the team to the change with a clear vision, strategy, motivation, measure of flexibility, and specification of immediate actions

  - Show change leadership by being serious, directive, overcommunicating, rallying supporters, receptive to feedback, and focused on quick wins

Resilient leadership has the power to deliver confidence and reassurance during times of uncertainty and confusion. Now that you understand how to build a resilient team, you have the foundation to grow your team in *Chapter 13*.

Further reading

Self-care:

- See Franklin Covey's *Habit 7: Sharpen the Saw* (https://www.franklincovey.com/habit-7/)

Hyping yourself up and calming down:

- *How to Pump Yourself Up Before a Presentation (or Calm Yourself Down)* by *Nancy Duarte* (https://hbr.org/2018/07/how-to-pump-yourself-up-before-a-presentation-or-calm-yourself-down)
- *How to Psych Yourself Up Before Your Next Big Thing* by *Vanessa Van Edwards* (https://www.scienceofpeople.com/psyched-up/)
- *The Debate of Power Posing Continues: Here's Where We Stand* by *Kim Elsesser* (https://www.forbes.com/sites/kimelsesser/2020/10/02/the-debate-on-power-posing-continues-heres-where-we-stand/)

Resiliency qualities and practices:

- *3 Practices That Set Resilient Teams Apart* by *Keith Ferrazzi* and *CeCe Morken* (https://hbr.org/2022/03/3-practices-that-set-resilient-teams-apart)
- *7 Strategies to Build a More Resilient Team* by *Keith Ferrazzi*, *Mary-Clare Race*, and *Alex Vincent* (https://hbr.org/2021/01/7-strategies-to-build-a-more-resilient-team)
- *The 4 Things Resilient Teams Do* by *Bradley Kirkman, Adam C. Stoverink, Sal Mistry*, and *Benson Rosen* (https://hbr.org/2019/07/the-4-things-resilient-teams-do)

13
Scaling Your Team

As demand increases for a product or business, engineering teams need to grow to meet that demand. **Scaling engineering teams** is the process of growing and expanding teams to meet new or changing needs in their work.

Scaling engineering teams requires, at a minimum, recruiting, evaluating candidates, interviewing candidates, making job offers, training new hires, adjusting divisions in responsibilities, and assessing the initial performance of new hires to ensure they are a good fit. Errors in judgment anywhere during this process can lead to costly setbacks, underperformance, and missed objectives. Problems in the hiring process may create friction that impedes teams for months at a time, draining their productivity. Engineering managers play a critical role in informing and guiding team scaling methods to avoid these pitfalls.

In this chapter, you will learn how to best approach scaling your engineering team. You'll learn how to navigate recruiting, assessing candidates, and making great hires. You'll learn how to onboard new hires with minimal disruption to your existing team and monitor the milestones in their performance. You'll learn what to do if the team grows so large that your previous practices are no longer sufficient.

By the end of this chapter, you will have the skills to navigate the ups and downs of scaling your team.

This chapter is organized into the following sections:

- Recruiting and hiring
- Onboarding new hires
- Managing a large team

This is a deep topic, so let's dive right into recruiting and hiring.

Recruiting and hiring

Recruiting and hiring are some of the most impactful activities we undertake as engineering managers. The new employees that we source and select for our teams will shape our work practices and end products for a long time to come, possibly well after we are no longer leading the team. They will make decisions that determine how we solve problems and eventually grow into the future leaders of our systems. Because of this, it is crucial that we make the best possible hires.

During recruiting and hiring for new positions, it is common for the demands of the process to dominate your schedule. This process will often leave you with little time for the rest of your responsibilities. These demands on your schedule, along with organizational pressure to fill open roles quickly, may encourage you to make hiring decisions too soon. While there may be pressure to rapidly fill open positions, there will be even more pressure if you end up with the wrong hires on your team. The time you spend trying to train and manage a bad hire while explaining slowdowns to your management team is much more costly than that of recruiting and hiring. Engineering managers save time and cost by doing whatever is necessary to make the best hires in the first place.

While recruiting and hiring are of crucial importance to get right, that doesn't mean that the process of doing so needs to be cumbersome. You may be in a highly structured work context where there are rigorous controls around hiring, or you may be in a lean environment where the hiring process is completely up to you to decide. Either way, you can orchestrate events to provide you and your leadership with confidence in your hiring decisions. This starts with setting your recruiting mindset and then proceeds through writing the job profile, building your hiring team, interviewing, assessing candidates, and marketing your positions.

Recruiting mindset

When beginning the hiring process, take a moment to manage yourself and consider what sort of team you intend to build. As engineering managers, we always want to make the best possible hires for our teams. We want strong programmers and **subject-matter experts** (**SMEs**) with relevant experience who can communicate well about the work and who fit the company culture. But hiring processes involve interpretation and personal judgment. It is common to end up falling back on hiring candidates that fit a narrow profile that we are comfortable with—that is, similar to each other in personality, educational background, communication style, and so on. This happens when we are seeking high confidence in our hiring decisions, and so similar candidate profiles end up feeling like a safe choice. For example, you might believe that the best candidates have degrees from certain schools, so you only hire those with that background. While those schools may provide an excellent education, when your team all have very similar perspectives, you have a narrower field of understanding as a group. The hidden risk in staffing with overly similar candidates is the same one we learned about in *Chapter 3*: trying to homogenize the team to a single profile. This is usually not effective. The best teams are well rounded and made up of complementary individual skills and thought processes. This is ideal for balancing day-to-day work needs since different types of tasks are often better handled by differently skilled and interested individuals. It is also true of exceptional product ideation, where you need different perspectives to come up with a wide set of ideas to inform the product.

The right approach to hiring is not to think about hiring star individuals, but instead about putting together a championship team. A championship team is one where team members play different positions in harmony with each other, each with different strengths, skills, personalities, and experiences to offer the team. Together, they have the ability to achieve a much wider set of objectives and learn from each other along the way. Individuals are important parts of your team, but work is performed in a team context, and the ability to contribute uniquely to that team can make the difference between good work and exceptional work.

As you begin the process of expanding your team, start by asking yourself, *Are there skills, perspectives, roles, backgrounds, or personalities my team is currently lacking? What might expand our capabilities as a group? What problems do we have that might be solved by introducing a new member?* Look beyond skills, and identify at the outset what qualities your new hire(s) might contribute to the improvement of overall team performance. For example, you might identify an area where your current team members may lack interest, such as documentation. Or, you might identify a desirable personality trait, such as skepticism or enthusiasm. Make a list of what you would like to add to your team and prioritize that list to refer to as you go through the hiring process. In *Chapter 16*, you will learn more techniques for identifying and prioritizing team member characteristics.

Once you have identified preferred qualities to add to your team, you can start to put together a thoughtful job profile.

Writing the job profile

To find what you are looking for, it helps to be specific. While the job description is your public-facing specification for a position, having a more concise "job profile" for internal use with your hiring team provides a more precise specification and a basic rubric for assessment. As shown in the following example, job profiles are concise, direct summaries of the hard skills, soft skills, and preferred qualities you are seeking in candidates for a particular role. The profile gives your recruiting team and interviewers a clear vision of the candidates you are seeking:

Job Profile: Senior Engineer, Identity Products

Seeking a senior engineer (SE3) who will work on identity products across legacy systems (Java) and some 3rd party integrations (JavaScript). These systems are not well documented or tested, so candidate must be highly resourceful, independent, pragmatic, and comfortable with ambiguity. Prefer strong systems design knowledge, TDD experience, and good at communicating with a range of stakeholders.

Figure 13.1 – Example job profile

You might be wondering, *Why can't I just use the job description and save myself some time?* The job profile and job description are not easily interchangeable because the documents serve different purposes. Job descriptions are written to cast a net that attracts desirable candidates. They provide information to job seekers about the position and company. They tend to be more general and may have euphemisms that describe work in favorable terms. They are also fairly long, to the point that they may not be read in detail by busy interviewers. On the other hand, job profiles are written to provide guidelines for candidate assessment. They provide interviewers with a frame of reference for deciding how well a candidate is matched for a particular position. They allow you to be direct and use company jargon without concern of dissuading candidates. A job profile is not an essential item in hiring, but it raises the efficiency and effectiveness of your hiring team by providing a concise reference they can incorporate into their busy days.

A good job profile should make it so that anyone from your recruiting, HR, cross-functional, and engineering team can understand what your idea of an ideal candidate is. It should outline your must-haves as well as the nice-to-haves. It should be concise enough that interviewers can read it quickly, but detailed enough that they know what you are looking for without having to ask follow-up questions.

You can include any information you want in your job profile, but it is good to start with at least technical skills, non-technical skills, and interests. You may want to also include personality, motivations, and working style. For each of these, it is important to not only list them but also quantify them.

Having an unambiguous job profile saves time in communicating about the position, but more importantly, it will help keep *you* focused on what you are looking for without getting distracted by candidates who are interesting but don't ultimately fulfill your immediate needs at hand. Once you have a job profile you are satisfied with, you can share it with your hiring team.

Building your hiring team

Except in the absolute leanest of start-up environments, recruiting and hiring are carried out with the cooperative efforts of multiple people. Depending on your workplace conventions, you may be responsible for choosing who you want to bring into the hiring process to help evaluate prospective recruits. The hiring team you assemble can have a big impact even if you ultimately have the final say as hiring manager. You probably will not go through with hiring a candidate that your colleagues find unsuitable or speak out strongly against. They also represent your team to the candidates who may be swayed for or against coming to work with you based on those conversations. Who you consult and how well they understand your needs are key considerations to making the best hires. Most engineer hiring processes will involve partnering with a recruiter, other engineers, and cross-functional team members.

Your recruiting partners

Recruiters source candidates and provide the first level of candidate screening. They are connected to the hiring market and can provide insight into where to find great candidates, which candidates expect from the process, and recruiting process statistics along the way. Depending on how things are set up at your company, sourcing and screening may be divided between different recruiting team members.

Regardless, these recruiting partners may be the most impactful to the success or failure of your recruiting and hiring process. If they don't understand what you are seeking in a candidate or they fail to hook those candidates with their pitch, you may find it impossible to get the talent level you are seeking.

To ensure a productive partnership with recruiting, make sure you remain in absolute lockstep with them. Don't assume they understand the nuance of what you are looking for in candidates. Be fully explicit with your expectations every step of the way.

Start by working with them on sourcing. Have your sourcing recruiter share all candidate profiles with you and give them detailed feedback on which profiles you think are good and what specific indicators make them good. Go through a couple of rounds of this with recruiting until you are confident that they are judging profiles accurately based on your criteria. Some may be resistant to going through this process with you because they believe they know how to source good candidates, but the reality is that there are no universal standards for candidate selection. If you rely on their judgment, you may end up weeding out valuable candidates based on criteria that you find irrelevant. They might place greater significance on details that you consider minor concerns. If your recruiter is resistant, tell them, *"I'm looking for something a little unorthodox with this group of hires, so I'd like to consider a wider pool of candidates and review them with you so that you get a sense of what I'm looking for."* This communicates that you respect their judgment and experience, but you want to take a specific approach to recruiting that warrants this calibration exercise.

Once you are happy with the sourcing, sync up with your recruiting partner on the screening. See if you can get a sense of how your recruiters are describing the position and what their conversations with candidates include. You can ask them about this directly or, if possible, it is even better to listen in on a couple of their screening calls. There are often aspects of the recruiter pitch that you will want to change for one reason or another. You might find that those conversations do not accurately portray your office culture, for example. Your recruiter should be a good representative of your team's vision and culture.

Your engineer partners

Your interview process will almost certainly require engineers from your team or other teams to help assess your candidates. With a strong job profile, working with engineer recruiting partners should be a simple process. For the engineers on your team, you probably know them well and have an understanding of where they may be stringent or lenient. Keep this in mind as you discuss candidates with them, since you may have differing opinions or goals.

The best technologists don't always make the best interviewers. How an interview is conducted can strongly influence how a candidate performs. If your workplace does not have a prescriptive approach to engineering interviews and it is up to you to determine, spend time with each of your engineering interviewers to get a sense of their style of interviewing. Try to choose the most appropriate interviewers for each position based on their personal style. For example, if you are hiring for a high-pressure position, you might choose a particularly rigorous and intense interviewer to give you confidence that the new hire can handle the day-to-day pressure.

Your cross-functional partners

Cross-functional team members are great additions to your hiring team. It is often a good idea to have engineering candidates interviewed by a project manager or product manager that works closely with your team. Other cross-functional roles that make sense to interview engineering candidates depend on the position responsibilities—for example having UI designers interview UI developer candidates. You can consider who this new hire will need to interact with the most and if it is appropriate to have those persons join your hiring team. Keep in mind that once they are on the team, it can be difficult to remove them, so make sure you only invite those who you are confident can add value to your hiring process.

Once you've identified cross-functional partners you'd like to work with, take a moment to set expectations with them. The job profile will establish a useful framing. From there, align with them further by sharing what aspects of their feedback you are most interested in or why specifically you had them in mind to join the hiring team. Communicating what you hope they bring to the hiring team gives them an area to focus on and helps them to help you.

With your hiring team assembled, your next step is to figure out how to assess your candidates.

Interviewing practices

Once your recruiting team has gathered a set of candidates for your open position, you can begin to assess those candidates. The primary method for gathering information to assess candidates for software positions is through interviews.

Many workplaces will have specific structures or practices you must follow for interviewing, but if yours does not, you can establish this for your team. Let's start with some guidelines for setting up your interview process and then learn approaches to technical and non-technical interviewing.

Setting up your interview process

Every company conducts interviews, but what makes a good interview process? A good interview process should be efficient and effective in gathering the right information about the candidates. An interview process should be the following:

- **Concise**: The faster you can move through the process and come to a decision, the better since candidates are often interviewing with multiple employers, and you may lose them at any point. Keep the process as speedy as possible to be respectful of their time and your time.

- **Consistent**: Uniform assessment produces the most fair and accurate results. When interviewers stray from uniform assessment, there is a much higher chance that unconscious biases will skew the results.

- **Progressive**: Have a means to weed out unsuitable candidates early on in the process, such as a technical screen phone call or test. Initial interviews should be simpler and easier to conduct, while later interviews are more intensive and take more of your team's time.

- **Finite**: You should establish the full interview circuit in advance. Having the process fully defined prepares the candidates to perform their best and makes sure that you get the most out of each interview since you are not counting on some future conversation.

- **Empathetic**: Keep in mind that you want candidates to be successful, so it is in your best interest to make sure they have an interview schedule with sufficient breaks, preparation time, and enough time to answer the questions they have.

Let's bring this together with an example. To design an interview process from scratch, you might decide your new position for an API engineer will have a three-round interview process. An **interview round** is a set of one or more interviews that a candidate must pass in order to advance to the next set of interviews. Our first round will be a technical screen phone call. Our second round will include technical interviews with senior API engineers and a data engineer. The final round will include non-technical interviews with the engineering manager and the project manager for the team. Our goal is to move candidates through all three rounds in a two-week period. Interviews will be scheduled for 50 minutes each with a 10-minute break at the end. The job profile and an interview scorecard will be provided to all interviewers.

Interview consistency can be a difficult ideal to live up to since you must aim for uniformity while avoiding a conversation that is overly formal, awkward, or robotic. We will learn some specific approaches to this in the next sections on technical and non-technical interviews.

Technical interviewing

Technical interviews assess a candidate's coding, problem-solving, and experience with specific tooling the job may require. There are many approaches to technical interviewing. The goal is to assess a candidate's technical ability as it relates to real day-to-day work. A candidate might be great at solving coding puzzles but poor at making real design decisions. Think about what skills you want to test for as you read through these interviewing methods:

- **Live coding** interviews provide the candidate with a **read-eval-print-loop** (REPL) or sandboxed code environment to solve a given problem in real time while the interviewer watches. The candidate must write functional code and often optimize and extend it to the interviewer's specifications. Live coding is useful in assessing the knowledge of particular languages, conventions, and understanding of algorithmic efficiency. On the other hand, it is criticized for being stressful and sometimes irrelevant to real work needs. Live coding interviews in practice end up testing not just technical ability but also how the candidate performs under pressure.

- **Take-home tests** provide candidates with a coding objective and allow them to complete it in their own time, submitting it to the interviewer for review. This practice provides a way to avoid skewing the results of candidates with the stress and pressure of live coding. The challenge with take-home tests is designing a problem that is meaningful and sufficiently demonstrates skill without being overly time-consuming. You must choose your questions carefully with this method so that you don't end up with candidates submitting their Google search results.

- **Code reading** interviews provide candidates with a piece of code and ask them to review it together with the interviewer. The candidate may be asked what the code does, if there are errors in it, how they might improve the code, or how they might extend its functionality. Code reading can be a great way to understand how a candidate thinks and their technical depth.

- **Systems design** interviews ask a candidate to design a feature or system verbally or with diagramming. Candidates are asked about data structures, storage, communication protocols, and performance. They typically make back-of-envelope estimates and then go into detail on one or more aspects of the design. These interviews are useful for evaluating the technical depth of senior candidates. Conduct these interviews with an empathetic tone so that they remain focused on real skills and not an ego-driven, overly competitive pursuit of cleverness or being "the smartest person in the room."

Depending on the position you are hiring for and your goals, you can use one or more of these interview types for technical assessment. Live coding and systems design are typically harder interviews, while take-home tests and code reading more easily lend themselves to any level of candidate or position. Ask yourself what is needed for your position. Consult your current team members to determine if your test is too hard or too easy. For any technical assessment, it is a good idea to include a basic problem with follow-up questions to provide an opportunity for candidates to distinguish themselves. Be sure to rotate your questions periodically since they may be shared online by some candidates.

Non-technical interviewing

In addition to technical skills, you will want to evaluate your candidates' non-technical skills. Non-technical interviews analyze a candidate across communication, time management, collaboration, interpersonal skills, leadership, and values. A simple approach to assessing non-technical skills is to focus on one (or two) of these broad areas per interview. This gives the interview focus and helps to avoid non-technical assessments devolving into popularity contests where, in the absence of specific evaluation criteria, an interviewer may end up preferring the candidate they enjoyed talking to the most.

Communication ability can be covered by your product manager—if you have one on your hiring team—since product managers work closely with engineers on requirements. If you don't have a product manager, communication is broadly applicable enough to be incorporated into any of your non-technical interviews.

Time management is a great focus area for a project manager if you have one on your hiring team. Otherwise, this can fit well into interviews conducted by a product manager or engineering manager.

Collaboration is a good choice for cross-functional partner interviews, such as UX designers. It can also be a good choice for peer engineering interviews.

Interpersonal skills and leadership are not always included in non-technical interviews, but they can provide another useful perspective. This can be incorporated into any of the discussions, but covering this yourself can help you appreciate the nuance of these skill sets.

Matching to company values can be considered throughout all interviews. It is a good approach to have each member of the hiring team provide feedback on how the candidate does or does not seem like a fit for the team's values. In the final round of interviews, the engineering manager can probe more deeply and ask follow-up questions that stem from the feedback already gathered.

Assessing candidates

After each interview, you collect feedback from the interviewers. The process of gathering feedback from your hiring team is deceptively time-consuming. Members of your hiring team have other meetings to attend, competing priorities, and urgent work to do that may cause them to put off giving their feedback. For many teams, gathering feedback becomes a bottleneck and obstacle to quickly moving candidates through the hiring pipeline.

When gathering candidate feedback, you want that feedback to be given soon and given with quantitative measures. Feedback is more useful when it is given soon, not just to shorten the overall assessment time but, more importantly, so that the feedback retains its fidelity. The longer it takes to capture the feedback from the interview, the more detail will be lost and the less useful the feedback will be. It is ideal to collect feedback immediately after the interview while it is fresh in the interviewer's memory. Feedback also needs to be quantified so that it can be compared across candidates. Interview feedback needs a consistent sense of scale for assessments to be meaningful and actionable. Hearing that one candidate is *great* and another is *fantastic* doesn't help to make a decision between candidates.

Feedback can be gathered verbally or in writing. To gather feedback verbally, you can schedule a quick post-interview debrief with the interviewers from that round. This has the advantage of reserving time for feedback so that you know you will receive the information promptly, but the disadvantages are that the interviewers may influence each other's opinions and it is harder to refer back to feedback given verbally. Instead or in addition, you may collect feedback from your interviewers in writing. This produces useful interview artifacts and maintains individual independence, but often takes much longer. Most established teams use written feedback for these benefits.

Measuring feedback

Whether you gather feedback verbally or in writing, one of the best things you can provide interviewers with is a scoring template for measuring candidate performance. This is sometimes called a candidate scorecard or grading rubric, as shown in *Figure 13.2*. A fixed scale helps your interviewers make more consistent and less biased decisions by directly comparing candidates on the same system of measurement. You will need a scale for technical interviews and one for non-technical interviews:

	Strong No	Soft No	Soft Yes	Strong Yes
Communication	Could not understand comms at all	Comms are disorganized, confusing, or contradictory	Comms are understandable	Comms are well organized and clear
Collaboration	No interest in or experience with collaboration	Expresses a desire to collaborate but lacks examples	Expresses a desire to collaborate and has experience	Expresses a desire to collaborate and has techniques
...				

Figure 13.2 – Example interview scorecard

Depending on the time and support you have available to you, you have the option of creating your own scorecard. Writing your own rubrics can be difficult and time-consuming but gives you total control over how candidates are evaluated. To save time, you may use publicly available candidate-grading rubrics. One source of scoring rubrics is interview-preparation websites, such as `techinterviewhandbook.org` and `tryexponent.com`, where they provide rubrics to help candidates prepare themselves, which you can adapt to your needs. There are a few examples given at the end of this chapter in the *Further reading* section.

Marketing your team

In addition to building and tuning your recruiting pipeline, engineering managers can recruit great new hires by marketing their teams. *Marketing* means to get the word out about your team and the work you are doing. Recruiters will handle basic job postings and maybe social media, but you can grow organic interest in your positions by taking some additional steps to raise awareness of your available positions, as follows:

- **Hosting community events**: Find a meetup or hackathon for relevant topics such as programming languages you use, and offer to host their events in your office (or otherwise provide resources for them).

- **Giving talks**: Give talks at meetups or conferences about the work you are doing and encourage your team to do so.

- **Writing blog posts**: Start an engineering blog for your company or contribute to external blogs to share some of the interesting challenges you are working on.

- **Partnerships**: Partner with community organizations, such as organizing volunteers to mentor at a young coder's organization in your community.

- **Sponsorships**: Sponsor a meetup, conference, or community organization, if possible. Talk to your leadership team about getting a small budget for this. Many smaller organizations are happy to receive any amount of sponsorship, so even $500 can be a meaningful contribution to your community.

- **Word of mouth**: Ask your current engineering team to refer engineers that they think would be a good addition if they feel comfortable doing so.

Adopting these practices helps boost awareness of your team and the work you are doing. The more awareness and interest you build in the tech community and within your own company, the more you can invert the recruiting process and encourage candidates to reach out to you.

Onboarding new hires

Onboarding is the process of setting up and preparing new hires to contribute work to the team. How you welcome your new hires sets the tone for their relationship with you and your organization. Onboarding experiences form a lasting impression. In the competitive landscape of engineer hiring, it is just as important to provide a smooth onboarding as it is a compelling recruitment experience. Good onboarding practices convey professionalism to your new hires and give them confidence that they have made the right decision in coming to work with you.

The process of onboarding a new software engineer typically includes providing hardware and software, access and permissions, orientation to systems and practices, and initial work tasks. It is ideal to be able to provide all of this information to your new hire within their first few days, along with making them feel welcome.

In addition to making a good first impression, your goal during onboarding is to minimize disruption to everyone involved. Good onboarding is timely and efficient and scales well to as many new hires as needed. To achieve this, leverage automation, provide training, and give structured performance objectives.

Leveraging automation

Onboarding scales best when manual tasks are reduced or eliminated. The more investment you have put into automating your team's day-to-day practices, the easier it will be to scale up your team's work to new contributors. Here are some examples of automations that simplify scaling up your team and onboarding new hires:

- Group-based access and permissions provide authorization to a group rather than to individuals so that you need only add a new user to the group to assign a wide array of authorizations
- Code-style conventions can be built into pre-commit hooks so that contributors automatically receive timely feedback on their code
- A **pull request** (**PR**) template describes what is expected from code reviews so that there is no need to explain
- Code quality conventions can be built into unit tests so that contributors receive feedback directly from the testing suite

Don't stop with these, but look for ways to automate any repetitive practices or processes. These automations offer benefits to team performance in general while also acting as **just-in-time** documentation that is especially useful for new hires. With this heavy lifting taken care of, you can focus your onboarding efforts on introducing new hires to the team and making them feel welcome.

Automation is incredibly helpful for granting access and providing code-level guidance, but for more general systems orientation, some training is necessary.

Providing new-hire training

To get your new hires ready to contribute to the team's work, they will need an understanding of your systems, practices, norms, and values. Documentation can be provided to give new hires the detailed information necessary for them to do their jobs.

Written documentation is particularly good for describing code, such as repositories and APIs. Repositories can use README files, and APIs can use tools such as Swagger.io. Documentation is also good for describing detailed expectations such as production support conventions. For documentation to be relied upon, it must be up to date and accurate. If your team cannot trust the documentation to be correct, they will end up seeking information from other engineers instead, disrupting the flow of work on the team.

Since accurate written documentation can sometimes be easier said than done, video documentation is a useful means to supplement what you have written. Video documentation reduces the barrier to both creating and consuming your docs. For less detailed explanations, short videos of a speaker explaining the material or a screencast demo are powerful means of conveying information in a way that is easy to understand and can be consumed broadly, on demand, and as often as needed. Video is also particularly good for recording screencasts demonstrating how to achieve specific tasks in a system. Videos can be captured in a video message archive such as Loom (loom.com), or in a simple video recording such as QuickTime Player on an Apple computer (**File** > **New Movie Recording**).

Text-based and video documentation provide asynchronous training resources that allow your new hires to come up to speed without inefficient knowledge transfer sessions disrupting your time or your team's time.

Along with this orientation, your new hires will need an understanding of what their objectives are in their first few months on your team.

Giving structured performance objectives

It helps to lay the groundwork for success by setting clear expectations for your new hires. Orient your new hires by providing them with a list of objectives for their first few months and a framework to check in on their progress along the way. This is sometimes called a 30/60/90 plan or just a 90-day plan. These objectives are intended to introduce your new hires to your team, practices, performance expectations, and the company as a whole.

Here are some things to include in your 90-day plan:

- Learning goals/expectations
- Contribution goals/expectations
- People to meet with or teams to be familiar with
- A schedule for checking in on progress
- A peer mentor, if desired

Each point of the plan should have enough objectives to be useful without overwhelming your new hires. It is common to have three to five objectives each for learning, contribution, and introductions. Schedule several check-ins during the first 90 days so that you know ahead of time if your new hire is facing obstacles in fulfilling the plan. Distribute objectives across 30-day, 60-day, and 90-day timeframes to help them work through the plan step by step.

Here is an example of a 90-day plan:

90 day plan for:

In 30 days

Learn: *Watch "welcome" videos; read values orientation docs*

Contribute: *push your first commit; have a pair programming session*

Meet: *introduce yourself to product manager and stakeholders*

In 60 days

...

Figure 13.3 – Example 90-day plan

When discussing the 90-day plan with your new hire, let them know that they can come to you at any point if they are having a problem with some aspect of the plan or need further guidance from you. Let them know ahead of time what happens if they fail to meet the plan schedule—if anything. Expressly communicating that you are there to support them and want them to be successful goes a long way to building their confidence in a new work environment.

Despite your best efforts, you might occasionally find yourself in a difficult situation where you regret a hiring decision.

How to handle a bad hire

Regardless of how good your hiring process is, you will on occasion make the wrong hire. Bad hires are statistically inevitable. It's okay to have the occasional false positive so long as you employ checks and balances to ensure that you address performance issues before they become a setback for your team. Follow these steps to minimize the impact of a bad hire:

1. Identify performance issues as soon as possible. Refine your 90-day onboarding template to include a reliable set of performance indicators that is representative of everything you need to see from your engineers. Use the plan along with general observation to alert you of potential problems ASAP.

2. Communicate performance issues to your manager and HR immediately. You will most likely be able to resolve performance issues with coaching, but in the rare cases where you are not, it is important to have set the groundwork for further action to be taken. Keep your leaders informed at every step.

3. Use radical candor and acknowledgment. Provide feedback to your new hire as soon as performance issues are observed. Give them the opportunity to acknowledge and explain their perspective on the situation.

4. Look for the root cause of performance issues to inform coaching. Take a blameless and empathetic approach to partnering with them on performance improvement. Let them know you are there to help and support them.

5. Decide within 90 days if performance challenges are coachable or not. If you have coached them consistently and they are still far away from the expected performance level, it might not be the right fit. If you need to part with your new hire amicably, it is typically easier to do so within the first 90 days.

Use these onboarding methods to scale your team as your needs grow. You may end up with a much larger team that needs you to reconsider some day-to-day practices.

Managing a large team

When your business grows and your team grows, you may eventually find yourself managing a much larger team. Growth on your team can be very exciting as it helps you grow your skill set as a manager. It can also be challenging when the processes and techniques that you have implemented to manage your team begin to lose some of their effectiveness.

When your team grows beyond a certain size, your methods of management will need to change. Generally speaking, the ideal number of direct reports is five to seven (see *Chapter 16* for more on this). When you exceed seven direct reports, it is very difficult to give the members of your team adequate time. This effect is somewhat modulated when your team grows slowly, so you have time to develop trust and shorthand with your team members before your team is larger, but it can be very pronounced when your team grows quickly.

As your team grows, it is also important to be aware that productivity does not scale linearly with team size. As you work with your leaders and stakeholders to plan for the growth of the engineering team, set expectations that your team's work output will increase, but it will not scale linearly. Team productivity growth is never linear because of the cost of increased coordination between team members. Even after onboarding and becoming acclimatized to the team's practices, communication and coordination across a large team have a cost. This is famously explained in Fred Brooks's *The Mythical Man-Month*, where he declares that adding manpower to a late software project only serves to make it later.

As your team grows, its management needs are changing. To wrangle your oversized team, you need a new team plan.

Short-term and long-term planning

Once you recognize that your team has exceeded a manageable size, it's time to make a new plan. Work on a short-term plan and a long-term plan to manage the situation. This does not have to be a formal document but can be captured in your notes and discussed with your manager.

The long-term plan is how you will eventually bring the team back to a manageable state. For example, you and your leadership might have a long-term plan to spin off a team from yours down the road or to introduce more structure to your team with managers reporting to you. To begin work on a long-term plan, you can start by talking to your manager to see if they agree that the current state is not ideal. You can ask your manager directly, "*What do you think about the size of the team?*" and go from there. Tell your manager what your point of view is on the team's expansion and how you would like to see it evolve.

Your short-term plan is what you will do to make the situation more manageable in the meantime. This plan should focus on how you will support your team's day-to-day needs. Specify what you will and will not take on personally, as well as how and at what cadence you will handle recurring meetings such as one-on-ones. The short-term plan is there to make sure that you and your team do not suffer while growing and transitioning to a new structure.

Delegating

Delegation is essential in large teams. As you work through your plans, consider which parts of your existing workload could potentially be handed over to members of your team. Senior members of your team can usually take on some of the complex tasks, but also consider how more junior members might be able to contribute. Delegation is a means to free up your time but also an opportunity for your team members to grow. Allowing your team to demonstrate their abilities and develop new skills can be an empowering and productive solution to a hectic team environment.

Focusing on people over projects

On a larger team, shift ownership of most or all projects to your senior engineers. Set up your engineers as leads on particular projects or product areas. This can be an opportunity for them to gain leadership experience while you shift your focus to supporting them.

Shifting your focus away from projects and toward people makes sense because your team members can lead projects but they can't support themselves, and they can't carry out the manager actions that you can. When the team scales, your focus should first be on the work that only you can do.

Preparing to switch contexts

With an oversized team, you can expect frequent context switching. *Context switching* means rapidly alternating between different areas of responsibility. As your team's responsibilities grow, you will be pulled from one area to the next with increasing speed and variety.

Manage your time rigorously to stay on top of different needs and workstreams. Take notes in each meeting, organize your notes at the end of each day, make to-do lists, and keep track of every commitment. Block out time on your calendar when needed. Keep a running tally of everything you have "in flight" to remind you of the range of work that may need your attention. Leverage technology as much as possible to reduce the mental burden of context switching, such as setting up reminders for yourself.

Making yourself available

As your team grows, team members have less access to you. Your time becomes more valuable and scarcer. Even if you don't believe that is the case, they may feel that they have less access to you or that they don't want to bother you when you are so busy. Since being disconnected from your team is never good, it helps to find deliberate ways to make yourself available to your team.

Try to have a few different means for team members to reach out to you. This can be as simple as keeping a standing agenda item for your weekly team meeting where you ask if anyone wants to raise anything with the group. You can set up professor-style "office hours" where you are available in a meeting room or on a video chat link for anyone who wants to talk. You can announce that you welcome anyone to reach out in your chat application's direct messages where you may not respond right away, but you will definitely get back to them when you have a moment. Choose the means that work for you and your schedule to allow your team to connect with you.

Summary

Scaling your team to meet new challenges is one of the most exciting times to be an engineering manager. Grow your team with care and intention for a lasting legacy within your organization.

First, focus on making the right hires to set your team up for success. Here's how you can do this:

- Start with the mindset of building a championship team where team members offer different skills and play different positions. Each new hire should expand the capabilities of the team.

- Write a job profile to provide a concise outline for yourself and others of what exactly you are seeking.

- Put together a hiring team of recruiters, engineers, and cross-functional partners to help you evaluate candidates. Deliberately set expectations with each of them so that they are crystal clear about what you are looking for.

- Plan your interview process from end to end, including technical and non-technical interviews.

- Provide interviewers with a scoring rubric to make assessments consistently and decisively.

- Spread the word about your team by writing, speaking, and getting involved in your communities of practice.

Next, make a great impression on your new hires with a smooth onboarding process, as follows:

- Simplify onboarding of new hires with automation. Keep your conventions and practices stored in code.

- Provide a mixture of written and video documentation to train new hires on your systems and processes.

- Give your new hires clear expectations and a structure to work toward them with a 90-day plan.

- Work closely with your new hires to determine as soon as possible if they are meeting expectations or not. Take action quickly if your new hire turns out to be a bad fit.

Lastly, as your team grows, adapt and evolve your team management practices, as follows:

- The ideal team size is five to seven direct reports. If your team exceeds that, pay extra attention to how that affects your ability to support them.

- Set expectations with your leadership that productivity does not scale linearly with new hires.

- When your team exceeds a manageable size, work on short- and long-term solutions to bring it back to manageability. Delegate more ownership, spend less time on projects, adopt rigorous time management practices, and take care to remain available to your team's needs.

Further reading

- *A Scorecard for Making Better Hiring Decisions* (https://hbr.org/2016/02/a-scorecard-for-making-better-hiring-decisions)

- *Coding interview rubrics* (https://www.techinterviewhandbook.org/coding-interview-rubrics/)

- *Behavioral interview rubrics* (`https://www.techinterviewhandbook.org/behavioral-interview-rubrics/`)

- *Google Coding Interview Rubric—An Inside Look* (`https://www.tryexponent.com/blog/google-coding-interview-rubric`)

- *How to document interview feedback for your hiring team* (`https://resources.workable.com/tutorial/document-interview-feedback`)

- *The Golden Rubric for Technical Interviews* (`https://medium.com/swlh/the-golden-rubric-for-technical-interviews-2f087ef2ba1`)

- *How to create a structured rubric for technical interviews* (`https://karat.com/blog/post/interview-engineering-how-to-create-a-structured-rubric-for-technical-interviews/`)

- *Time Management for Engineering Managers* (`https://blog.xendit.engineer/time-management-for-engineering-managers-a3c95f399e3`)

14

Changing Priorities, Company Pivots, and Reorgs

In *Part 4* of this book, you have learned how to increase resilience in engineering teams and how to manage scenarios where you must scale up your team. In addition to changes in *size*, teams experience organizational *shifts*, moving from one thing to the next. Shifting occurs at the micro level—tasks and projects—as well as at the macro level—product and company. To achieve continued success, engineering managers must guide their teams through these shifts.

This chapter introduces concepts and techniques to help teams shift from one thing to the next. We will start with a survey of prioritization methods and see how to select the right method. Then, we will look at how you can manage situations where priorities constantly change. Next, you will learn how you can protect and promote your teams during major organizational changes such as company pivots and team **reorganizations** (**reorgs**). By the end of this chapter, you will know how to take a structured approach to lead your team through substantial changes while maintaining their trust, goodwill, and momentum.

This chapter is divided into the following sections:

- Prioritization
- Managing changes in priorities
- When objectives or structures change

Let's start by building upon what you learned in *Chapter 5* by taking a deeper look at prioritization.

Prioritization

As we plan and negotiate complex workloads, we attempt to solve decision problems with prioritization. It aims to answer questions such as, what features should we build? What should we do first? What tasks can we defer if we run out of time or resources? In *Chapter 5*, we introduced prioritization as an important step in overall project planning, but we did not broach areas such as the different methods, the pitfalls of prioritization, and how to choose the right method.

Prioritization in itself is a deep topic, with scores of competing approaches and decades of research examining its conditions and practices. While we cannot cover all that there is to know about prioritization, it is useful to understand some of the trade-offs as a foundation for managing shifting priorities. With that in mind, let's start with a survey of common prioritization methods.

Methods of prioritization

Methods of prioritization range from simple to complex. Regardless of the methodology, when put into practice, prioritization methods often suffer from similar pain points—difficulty in handling large sets of requirements and difficulty with prioritization updates (re-prioritization). Since software projects commonly have large feature sets and undergo frequent re-prioritization, it is our goal to find the best methodology for the given and additional contexts within our work. With this in mind, let's look at the Eisenhower method, numerical prioritization, stack ranking, and dot voting.

Eisenhower method

The Eisenhower method is a simple means of prioritizing that comes from a quote attributed to Dwight D. Eisenhower: "*I have two kinds of problems, the urgent and the important. The urgent are not important, and the important are never urgent.*" *Figure 14.1* shows the Eisenhower matrix:

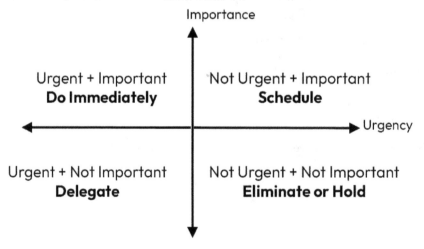

Figure 14.1: The Eisenhower matrix

This method of prioritization plots work across quadrants with axes of **Urgency** and **Importance**. This allows structured consideration of tasks and helps to avoid scenarios where urgent work overshadows important work. This method is used more often to plan tasks and day-to-day work rather than project requirements.

Numerical assignment

Numerical assignment or numerical prioritization is one of the most widely used methods of prioritization and is what we described in *Chapter 5*. Numerical assignment refers to establishing numbered groups of priorities, such as *priority-1*, *priority-2*, *priority-3*, and so on. Such prioritization is most successful when the numbers correspond to understood levels that are well defined to avoid differing interpretations such as "*critical: must be implemented prior to launch*" and "*high: should be implemented prior to launch.*" Further, it is practical to limit how many requirements can be assigned to each priority group, such as saying "critical" can have no more than 30% of the total requirements. Finally, it is helpful to provide reference priorities to demonstrate how prioritization groups should be applied to new requirements—for example, capturing "_____ *is an example of a critical requirement because…*" in your documentation.

Numerical prioritization also includes specific group assignment methods, such as **MoSCoW prioritization** where the four priority groups are *must-have*, *should-have*, *could-have*, and *won't-have*.

This method of prioritization scales reasonably well but does not represent differing stakeholder opinions if there are multiple stakeholders. It can be a challenge to gain consensus on priority across a varied stakeholder group.

Stack ranking

Stack ranking or ranking is a common method of prioritization where requirements are placed in priority order from 1 to *n* requirements. Stack ranking is scalable and works best when requirements are captured such that they are completely independent of one another. It is beneficial to have an exact priority order for an entire body of work that can be followed sequentially. However, this method is criticized for lacking any indication of the magnitude of priority difference between items, as well as lacking representation of dependencies. If there are multiple stakeholders, this method requires gaining consensus across the stakeholder group, which can prove difficult.

Dot voting

Dot voting is a simple method of prioritization where stakeholders "vote" on requirements by placing a marker on the requirements that are a priority for them. Voting is sometimes done in person using Post-It notes or stickers, but it can be done virtually on a digital board just as easily. In a multi-stakeholder context, this method offers a convenient visual representation of which features are important to the highest number of stakeholders. Variations of dot voting include the **hundred-dollar test** (sometimes called *buy-a-feature*), where stakeholders are each given some amount of (fictional) priority dollars to assign to their choices as desired, and **top ten prioritization**, where each stakeholder chooses their top 10 priorities (or top *n* priorities). The hundred-dollar test gives more flexibility to stakeholders to weight their priorities according to their needs, but it makes it difficult to tell whether a priority is widely needed or only important to a single stakeholder. All dot voting methods have been criticized for encouraging bandwagoning, where influential stakeholders sway the opinions of others, making the results less representative.

Choosing a prioritization method

If you are in a position to choose from different prioritization methods, the prevailing wisdom is to choose the simplest form of prioritization that meets your needs. The primary factors determining your needs are, typically, the number of stakeholders, the number of requirements to be prioritized, and available time. The likelihood of re-prioritization can also be considered. *Figure 14.2* shows prioritization methods ranked by simplicity:

Technique	Scale	Granularity	Sophistication
AHP	Ratio	Fine	Very Complex
Hundred-dollar test	Ratio	Fine	Complex
Ranking	Ordinal	Medium	Easy
Numerical Assignment	Ordinal	Coarse	Very Easy
Top-ten	-	Extremely Coarse	Extremely Easy

Figure 14.2: Prioritization methods ranked by simplicity (Berander and Andrews, 2005)

Numerical and stack ranking work best when stakeholders are few or are well aligned to begin with. Dot voting methods work well to gain consensus from a diverse set of stakeholders. Eisenhower prioritization is typically applied to day planning and time management, but it can be useful to sanity-check your team's workload and ensure that the work you do is important and not just urgent.

Each of these methods has shortcomings and can become a burden to coordinate when requirement lists stretch into the hundreds and beyond. For this reason, the ideal choice is to do the least prioritization possible while still being able to make informed decisions.

Now that we know a few approaches to prioritization, let's look at the process of re-prioritization.

Managing changes in priorities

As new information becomes available, priorities change. Generally, it is a good thing to have flexibility in prioritization to respond to changes and course-correct. However, sometimes, the degree of change may become excessive, and the source of change can feel less like "new information" and more like "the inability to make up one's mind." Excessive changes in priorities may become frustrating to engineering teams when they cannot get clarity on what they are meant to be working on. When priority changes become more representative of a lack of commitment or lack of focus, this can signal that it is time for an engineering manager to investigate.

Churn in software development refers to how often something is overwritten or changed. To get a handle on priority churn, start by looking at the context and the prioritization dynamics, and then use that information to determine how to address the situation.

Prioritization context

To understand prioritization challenges, start by looking at the context. Though priority churn is common, its causes are not always the same. You can start with the following questions to understand the context:

- *Who are you building for?* In other words, who are your first-order and second-order customers? For example, you might build a web product for a mass market where your primary customers are your users and your secondary customers are advertisers. Alternatively, you may build software for a client where the client's wishes are most important, but you need to consider end users as well. The customers determine how you might look at prioritization.

- *How many stakeholders do you have?* Do you have one or two powerful stakeholders in your organization, or is there a consortium of stakeholders to whom you must cater?

- *Is the software delivery fixed or iterative?* In other words, how frequently are you able to revise and ship a new version of the end product of your work? Consider both the medium (for example, web versus native application release cycles) and the delivery tooling (that is, software delivery pipeline maturity).

This information provides a basis to frame your solutions, but before solutioning, assess the situational dynamics.

Prioritization dynamics

Next, to understand the prioritization challenges, look at the organizational dynamics that are in play. These dynamics may require some effort to uncover or they may be painfully obvious. Either way, determine the following.

Why are priorities changing?

Figure out where your priority churn is coming from. There can be many underlying causes of frequent priority changes. It might be that your stakeholders compete against one another, continuously battle, and escalate their individual priorities, leading to frequent changes. It might be that stakeholders don't really know what they want and change their minds frequently. It might be that stakeholders are impatient and have a hard time waiting for results before suggesting a new course of action. Whatever the case may be, try to get to the root of why priority churn is high by interviewing team members and the stakeholders themselves.

How does priority churn impact the team?

Since priority changes aren't inherently bad, it's useful to construct a full picture of the situation by taking some introspective measures. If priority changes cause problems, what specifically are those problems? It might be that priority changes make it difficult to maintain planning cycles, such as agile sprints. It might be that team members are interrupted in the middle of a unit of work, leading to unreleased feature work becoming stale. It might be that team members feel frustrated that their work is disrupted. It might be that team members feel that they never finish anything substantial and gain satisfaction from their accomplishments. Observe how priority churn impacts your team and talk to them to determine the extent of the challenges they face.

Once you have assessed the prioritization factors in effect, you can work toward solutions.

Prioritization solutions

With what you have learned so far in this chapter, there may be an obvious solution to your prioritization problems. If your problems appear to stem from using the wrong prioritization method for your workplace context, you might be able to get a quick win by switching your prioritization to a more suitable method. For example, switching from a method that works better with a single stakeholder (stack ranking) to a method that works better with many stakeholders (dot voting) might be the solution to your problems. Otherwise, if prioritization churn persists, the next step is to address the dynamics you uncovered.

Given the common scenarios described in the *Prioritization dynamics* section, the following approaches can help reduce prioritization churn:

- **When stakeholders compete with one another**: Try switching to a multi-stakeholder prioritization method, such as dot voting, or incorporate stakeholders into a periodic planning cycle, such as quarterly planning, where each stakeholder gets some of their requirements included and knows what they can expect during the upcoming quarter.

- **When stakeholders don't know what they want**: This is usually best solved by partnering with stakeholders to help them figure out what it is that they really care about. Often, they have a business metric they want to impact but flip-flop on the right approach to do so; find out what that metric is and help them determine the best approach.

- **When stakeholders are impatient**: This can be addressed by involving your stakeholders more closely in the delivery process so that they can see the progress being made. Communicate with stakeholders frequently and capture results via analytics to share with them.

You can further address the effects of prioritization churn on your engineering team as follows:

- If planning cycles are a problem, you can switch to a methodology that doesn't rely on planning cycles, such as **kanban**, which re-prioritizes top items by continuously stack ranking individual work units. This requires maintaining a well-groomed backlog of work.

- If work is disrupted, leaving projects partially complete, you may find that breaking work down into small units that you can release in a disabled state via feature flags allows you to more easily switch between work on different features.

> **Feature flags**
>
> Feature flags, or feature toggles, are a deployment method where configuration is used to enable and disable different functionalities. This technique allows you to deploy code without making it available to users.

- If team members are frustrated by changes and lack of continuity, you can work on building up their resilience, as described in *Chapter 13*. Find a way to give the work meaning despite the changes, and help your team members to see how their work makes a difference. Keep track of your team members' assignments, and take advantage of opportunities to give them work that helps them achieve their individual goals and career continuity.

Small changes you make in your process and communication can improve situations with shifting priorities, but they may not be able to completely resolve priority churn. If possible, you are often better off minimizing granular prioritization in favor of more user testing.

User testing

While prioritization attempts to solve decision problems, those decisions are routinely built on top of uncertain information. Roadmaps are often created based on what product teams *believe* their users want, without always having a deep understanding of those user dynamics. When we assign priorities and make decisions based on what we believe people want instead of knowing what they want, there is much more room for interpretation, disagreement, and churn. When we are able to see unequivocally what our users are interested in and respond to, making product decisions becomes considerably easier. This is why user testing is so beneficial.

User testing, also called A/B testing, multivariate testing, or experimentation, is the practice of releasing multiple concurrent versions of software and dividing traffic between the versions in order to collect usage data and determine which features or attributes are most successful. User testing is relevant to decision making primarily in product contexts, where you build for end users rather than clients or businesses. In a user-centric setting, experimentation can reduce prioritization struggles by providing real data to inform decisions. Stakeholders can put forth data-driven hypotheses for testing, and the development team can release a **minimum viable product** (**MVP**) to gauge and proceed based on real user feedback.

User testing encourages product engineering teams to take more ownership of features. When you can perform experiments to see what drives business metrics, then the product team can take an active role in making the product decisions that move those metrics. This way, stakeholders can communicate the business metric improvement they want to see and allow the product team more creativity in finding ways to improve those metrics. With this approach, stakeholder prioritization becomes much more high-level and churn drastically reduces.

Now that you know how to handle changing priorities, let's look at bigger changes in our objectives and structures.

Managing changes in objectives or structures

Pivot is a way to refer to a significant change in your product focus or objectives. *Reorg* is a common abbreviated term used to describe the events that reshape the structure and hierarchy of teams. As the needs of your business change, you might be faced with a pivot, a reorg, or both.

Organizational changes occur in response to various scenarios, including mergers and acquisitions, changing market conditions, and striving to improve business operations. During these changes, engineering managers have the responsibility of easing the difficulties their teams face and supporting them through the changes.

Engineers are often skeptical of organizational changes. If you have had enough time with your team to build resilience and positive team emergent states, such as trust and psychological safety, you will be in a better position to help them weather the coming changes. In addition to that, pursue the following steps.

Understand the changes

During organizational change, the first thing you can do to help your team and yourself is to understand the purpose of the changes you are facing. Pay attention during announcements, ask questions, and reach out to senior leaders you know to get the most complete picture you can. Knowing the intentions of an upcoming organizational change helps you to prepare your team. It helps you to answer their questions, contextualize changes, and give work meaning. You may not be able to obtain clear information right away, and you may not have all the answers to your team's questions, but it is helpful to make an effort to gather the information and inform your team that you will follow up on their questions. Being able to clearly explain why organizational changes have happened can make the difference between acceptance and resistance from your team members.

Understand your team

The next thing you can do to support your team during organizational change is to make sure you are aware with your team members' desires. As their manager, you may have opportunities to help them reach their individual goals, but only if you know what those goals are. Make sure you have sufficient time with them for two-way communication. Talk to each team member in your one-on-one meetings and ensure that you understand their priorities and preferences. Tell them you don't know for sure whether you will have input into changes, but if you do, you want to be sure you are aware of their priorities. Ask whether or not they are interested in changing teams or working on a different project or technology. Ask them whether there is anything in particular they hope to get out of the changes. Ask them whether there is anything they know they definitely don't want to do. Reiterate that you don't know if you will have an opportunity to use this information, but you want to be prepared if an opportunity comes up.

Secondly, make sure you understand the strengths and weaknesses of your current structure and work processes. As changes unfold and opportunities arise, it is incredibly useful to have a clear point of view of what you want to preserve and what you want to overhaul.

Be a guide and an advocate

Once you understand the purpose of the change and the desires of your team members, the next step is to steward those aims as a guide and an advocate. Translate the changes in terms your team can understand, and steer them toward the right actions while they focus on the work to be done.

In meetings with your manager, advocate for your team by positioning them for the outcomes they seek (that you are supportive of). If possible, use self-selection by engineers to organize and assign teams. Emerging research indicates that self-selection can produce better outcomes and higher motivation during organizational change (you can read more at `https://pragprog.com/titles/mmteams/creating-great-teams/`).

If you have an opportunity to meet with senior leaders in your company, use the opportunity to advocate for your team in a context appropriate way. For example, in a private meeting, you might advocate directly for your individual team members, while in a group meeting, you could make general suggestions to benefit your team.

Understand the directive

As organizational changes are rolled out, you might find yourself with a new product direction, a new role, or even a new team. At this stage, your most important responsibility is to develop and understand your team's new directive. Dig into your area of responsibility to discern how it may differ from what it was in the past. For example, after an organizational change, you might have the same product responsibility, but now you need to work with a different process and stakeholder hierarchy. Ask questions to your manager and leadership until you are confident that you understand what is expected of your team. Also, make sure you understand any phased releasing of the organizational changes that may impact your team.

Align the team and stakeholders

Once you understand the new directive, you can start to align your team with it. Think about how to frame an explanation for your team to give the work meaning. Identify and communicate any advantages you see in the new direction. Provide your team with a clear structure and expectations—for example, telling them that you will begin working with the new process after the current planning period ends on a given date. Overcommunicate changes and expectations to help those changes sink in. This is a good time to adopt a more authoritative leadership style, as you learned in the *Commander and servant* section of *Chapter 2*.

If team members change managers, hold **manager hand-off meetings**. Sometimes called a *one-on-one-on-one*, manager hand-off is an opportunity to pass the baton between managers to ease the transition and minimize the loss of career momentum for engineers. The agenda should include individual performance (strengths and growth areas) as well as agreements between the engineer and the former manager. These meetings work well when led by the former manager, walking the new manager through the specifics and giving the engineer room to contribute and ask questions.

Build alliances

Along with giving your team clear direction, seek out your new cross-functional partners and stakeholders. If there are new faces in your orbit, familiarize yourself with them. Start to get to know them and build alliances with them. It will take some time before new practices become habitual, so having close partners to help you adjust to changes and work through problems can be a big help. Work to develop trust and mutual support with them so that you are prepared to support each other and navigate the changes together.

Build momentum

The last big step to take during organizational change is to reinforce changes and build momentum by putting some quick wins under your belt. Quick wins are small goals or *low-hanging fruit* that can be rapidly attained. Achieving goals helps the team to build confidence in their abilities and velocity in their work. Developing belief in the team early supports them in overcoming further challenges and increases change acceptance. This momentum also helps the team move through the *four stages of a team* more rapidly (see *Chapter 9*).

Summary

In this chapter, you learned techniques to manage organizational change. Use the information in this chapter to help you negotiate challenging scenarios with shifting expectations for your team. Remember the following:

- The term *prioritization* encompasses a range of methods that help you to make decisions about what work to do by providing supportive frameworks. Some of the most commonly use prioritization methods are as follows:

  - The Eisenhower method, which ranks work according to urgency and importance

  - Numerical assignment, which assigns work to ranked priority groups

  - Stack ranking, which ranks individual requirements against each other

  - Dot voting, which collects and tallies individual votes for requirements

- As new information becomes available, expect your priorities to change to accommodate the latest information.

- When priorities change excessively, we refer to it as high *priority churn* and address it in the following ways:

  - Determine why the churn is happening and address the root causes

  - Determine how the churn impacts the engineering team and address those problematic effects

  - If possible, incorporate user testing to provide direct user feedback and inform the prioritization process

  - Work with stakeholders to prioritize outcomes in business metrics rather than specific features that may or may not impact those metrics

- Common organizational changes include *pivots* (new directions) and *reorgs* (new hierarchies).

- When your team faces a pivot or reorg, focus on the following actions:

  - Understand the changes—why they are happening, and what they are meant to accomplish?

  - Understand your team—what do individual team members want for themselves?

  - Be a guide and an advocate, supporting and campaigning on behalf of your team

  - Understand the new direction—what are you meant to do, and how are you meant to do it?

  - Align your team with expectations, providing strong guidance

  - Build alliances with new stakeholders and cross-functional partners

  - Build momentum by setting up and achieving quick wins

With that, you have the foundation to handle the shifts that your team might face on your journey together. Next, in *Part 5* of this book, you will learn how to support your team over a longer period, starting with how to retain talent.

Further reading

- *Berander, Patrik*, and *Anneliese Andrews*. *"Requirements prioritization." Engineering and managing software requirements* (2005): 69–94 (`https://link.springer.com/chapter/10.1007/3-540-28244-0_4`)

- *Mamoli, Sandy*, and *David Mole. Creating Great Teams: How Self-selection Lets People Excel.* Pragmatic Bookshelf, 2015 (`https://pragprog.com/titles/mmteams/creating-great-teams/`)

- *Recardo, Ronald J., and Kleigh Heather. "Ten best practices for restructuring the organization."* Global Business and Organizational Excellence 32.2 (2013): 23–37 (`https://doi.org/10.1002/joe.21470`)

- *Watkins, Michael D. The first 90 days, updated and expanded: proven strategies for getting up to speed faster and smarter.* Harvard Business Review Press, 2013 (`https://hbr.org/books/watkins`)

Part 5:
Long-Term Strategies

In this part, we will introduce more advanced topics for you to further explore. These chapters cover additional techniques to strengthen and support teams in the long run. You will learn ways to retain your team members, ways to structure your team, and answers to a few lingering questions you may still have.

This part has the following chapters:

- *Chapter 15, Retaining Talent*
- *Chapter 16, Team Design and More*

15

Retaining Talent

Once you are comfortable with the day-to-day work of leading your team, as an engineering manager, you begin to look more to the future. As you implement the strategies from this book and your leadership efforts evolve from big changes to small improvements, your focus shifts to how to retain what you have built. In *Part 5* of the book, we will look at the work of engineering managers over a longer time horizon and begin by exploring long-term strategies to retain talent.

As software development is a type of knowledge work, much of the value we add to our projects stems from the collective knowledge of our team members. For that reason, as managers, our task to preserve what we build necessitates retaining these valuable individuals within our teams. Great engineering managers need to know how to retain their team's talent. Compensation may initially hook talented engineers, but it is often not enough to retain them when other offers come along.

In this chapter, you will learn tactics to retain engineers on your team over time. You will start by learning why talent retention matters. Next, you will learn how to arrange your team members' environments, careers, and relationships to maximize their job satisfaction. Finally, you will learn where talent retention can go wrong. By the end of this chapter, you will understand how to keep your engineering team members engaged and thriving.

This chapter is organized into the following sections:

- Why should you retain talent?
- What does it take to retain talent?
- The pitfalls of retention

Let's start with why talent retention matters.

Why should you retain talent?

The **turnover rate** refers to how many positions on a team are "turned over" or vacated and then rehired. According to `Linkedin.com`, their data shows a turnover rate for technology teams at around 10–13% per year. When positions turn over, engineering managers have the ability to hire new team members and coach them to success, but there are many benefits to retaining the talent you already have on your team as follows:

- **Cost**: Hiring and onboarding new staff is expensive. According to the **Society for Human Resource Management (SHRM)**, each employee departure can cost between ⅓ to ⅔ of the employee's annual earnings. Both in terms of resources and time, restaffing requires considerable effort. As you learned in *Chapter 13*, you can expect a long list of activities, ranging from writing job descriptions to scheduling interviews and wrangling interviewers. There is also a bottom line cost for job postings, recruiter fees, referral fees, and other costs. By retaining the team members you already have, you save your company considerable time and money spent on rehiring.

- **Productivity**: Retaining talent is beneficial because it also retains productivity. New hires may take a year or longer to be as proficient as tenured members of the team (`https://www.lingolive.com/blog/cost-of-turnover-for-software-engineers/`). Your existing team members know your code base and probably have written significant portions of it. No matter how strong a new engineer is, it takes time to come up to speed on new systems and team conventions. That time is lost when your positions turn over, and it is likely to take months to reach the same level of productivity you had before. Additionally, there is lost productivity from organizational know-how and relationships, since new hires face a learning curve in navigating the organization and their colleagues.

- **Reputation**: While more intangible than cost and productivity, it is also beneficial to work toward retaining talent because, when you do so, you create a more desirable environment in general. Talent retention efforts produce greater numbers of team members who are proponents of your workplace. Great workplaces benefit from better reputations and word-of-mouth advertising so that when you do need to hire, you can do so more quickly.

Now that you understand why it is worthwhile to retain talent, let's move on to how we can do so.

What does it take to retain talent?

Throughout this book, you learned methods to create a desirable work experience. You have learned how to give work meaning, how to give yourself and your team a sense of purpose, and how to facilitate productive team emergent states that improve attitudes and work outcomes. In many cases, applying the techniques you learned in earlier chapters will be enough to keep your team engaged and retain them.

Beyond providing a thoughtfully curated work experience, engineering managers can increase confidence in retaining their teams by striving to increase workplace satisfaction. In addition to your work on general leadership practices, focusing on satisfaction is an opportunity to examine how your leadership is received and what might be lacking. There is no way to guarantee you retain talent in all circumstances, but working to assure satisfaction helps to avoid surprises in discovering that an engineer on your team found some aspect of their work experience to be lacking. For comprehensiveness, we will follow a structured approach and look at satisfaction with the work environment, with individual growth, with management, and with the company as a whole.

Satisfaction with the work environment

In your efforts to ensure your team members' satisfaction, start by confirming the workplace environment is what it needs to be. *Environment* in this sense is not only the atmosphere but also the norms, behaviors, and practices.

You may be familiar with Maslow's hierarchy of needs. Maslow theorized that our individual needs form a pyramid that stacks from the bottom where the needs at each level must be met before our needs move up to the next level. This hierarchy starts with physiological needs and then moves upwards through safety, social, esteem, cognitive, aesthetic, self-actualization, and transcendence, as shown in *Figure 15.1*:

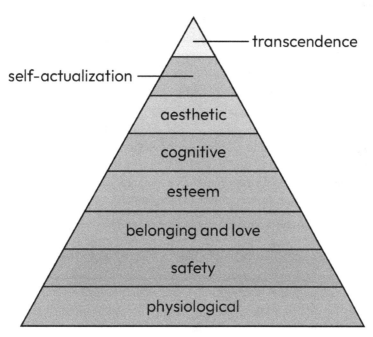

Figure 15.1 – Maslow's hierarchy of needs

What Maslow's hierarchy can teach us as engineering managers is that satisfaction is not a static point but a moving target, where we achieve milestones in succession. Looking at workplace environment satisfaction through the lens of Maslow's hierarchy, we can identify the following areas of interest in each layer:

- **Physiological** needs include things such as food, water, and sleep, so they do not need to be considered in the context of the workplace environment except to state the obvious—your team members need to take care of these basic needs for themselves.

- **Safety** needs include job security and financial security. These are foundational to satisfaction with the workplace environment. For example, if a member of your team works with you on a temporary contract basis, they may not have the job security to be satisfied with their position. Conversely, once this level of financial need is met by a salary, other factors become more important in retaining them.

- The need for **belonging** is our desire for acceptance within social groups. In terms of the work environment, this relies on identifying as a team, as discussed in *Chapter 9*, as well as feeling accepted by their team. To produce a satisfying environment, make sure each team member feels like they belong. Be inclusive by accommodating and integrating different viewpoints within your team.

- **Esteem** needs refer to the desire for self-respect and mutual respect among team members. In *Chapter 6*, we learned an approach to instilling pride in the work being done to build a virtuous cycle of self-respect. Then, in *Chapter 9*, we learned how to facilitate motivation by providing autonomy, mastery, and purpose. These are the building blocks of esteem in the workplace environment. Create a workplace environment that allows team members to develop their individual confidence to produce greater satisfaction for them.

- **Cognitive** needs encompass the means to intellectual satisfaction, such as creativity, functional mastery, and overcoming challenges with novel solutions. In the workplace environment, engineering managers can encourage intellectual satisfaction by building creativity, curiosity, and challenge-readiness into the team climate. Demonstrate these qualities in your own behavior and reinforce them in others through recognition. Hold exercises that allow your team to flex their intellectual muscles, such as brainstorming sessions.

- **Aesthetic** needs refer to the desire to have pleasing experiences, elegance, and beauty in one's life. Engineers may experience this through a well-designed workplace, through an elegant bit of code, or simply by taking a break and going on a nature walk. There is not much to do as an engineering manager to encourage this, except to acknowledge that the need for beauty is genuine and do what you can to facilitate it. For example, an engineer might choose to take a position partly because they love the design of the office space. You may rarely have control over the aesthetic aspects of your work environment, but when you have a chance to influence them, it is a good idea to do so.

- **Self-actualization** and **transcendence** are a later addition in Maslow's work to represent individual fulfillment and spiritual fulfillment respectively. The main influence in the work environment you provide that encourages self-actualization is the set of productive team emergent states you learned in *Chapter 9*. To further this, give your team members satisfaction with growth and opportunities.

Satisfaction with growth and opportunities

To work toward retaining talent, the next area to look at is your team members' satisfaction with their access to growth and opportunities. For this, let's revisit three of Maslow's needs—esteem, cognitive, and self-actualization.

Esteem

Growth and opportunities are essential to the virtuous cycle of pride and self-respect you learned in *Chapter 6*. Developing pride and self-respect relies on accomplishments and achievements provided by your work. As an engineering manager, you can provide for growth and opportunities structurally as well as individually.

First, affirm that your team members have access to structures for professional growth. In other words, make sure that processes and practices exist for growth within the team. This includes the following:

- Is career advancement information readily available to them?
- Do they understand the requirements for career advancement?
- Do career paths exist for management as well as individual contribution?

For those career-focused members of your team, growth structures provide a system to help them get what they want out of their work experience. Clear paths to advancement are essential to job satisfaction for many people.

If your workplace does not have clear expectations for career advancement and career paths, advocate for it with your manager and senior leadership. Career paths should include positions, position expectations, and documentation around advancement (refer to the *Setting expectations* section in *Chapter 8* for an example of career advancement documentation). If you need to create career-pathing documentation yourself, there are many publicly available examples of software career paths or ladders to help you create your version, such as `progression.fyi`.

To provide for esteem individually, you can connect with each member of your team to confirm that they feel they have sufficient access to growth and opportunities. Since everyone is different, it is a good idea to find out what aspects of growth are important to each of your team members. Some may be driven by career advancement, while others are only interested in skill development, and others may desire growth through autonomy and the opportunity to take ownership of more impactful projects. Figure out what matters to each member of your team so that you can tailor your efforts toward each of them.

Cognitive

A literature review on motivating software engineers found that the single most common characteristic attributed to engineers is a *growth orientation* or intrinsic desire to be challenged and learn new skills (DOI: `10.1016/j.infsof.2007.09.004`). The vast majority of engineers are interested in opportunities to develop subject matter expertise. Distinct from the need for esteem and self-respect, engineers want to be challenged purely to grow their knowledge and skill sets. Much like the satisfaction of solving a puzzle, cognitive needs are their own reward.

To provide growth and opportunities for cognitive needs, help your engineers attain excellence in their specializations. Assess their level of access to resources within your company that help them to continuously develop skills and challenge themselves, including the following:

- Is skill development or training available to them?

- Do they have access to conferences?

- Is challenging work available to them?

- Do they have access to colleagues with the same interest areas?

Advocate for team members' access to resources to help them grow and promote engineering excellence. Many companies will provide some educational budget to spend on training or licenses to online course content. Give out technical books as prizes or build an in-house library for the team. Find out what conferences are relevant to your work, and consider sending team members to them.

Find ways to provide sufficiently challenging work for your team. Carve out assignments for each team member that helps them to deepen their knowledge and build upon their skills. This doesn't need to be 100% of their work, but you can have regular time devoted to it. If you are struggling to find enriching work, you can arrange for a hack day or 20% time to allow them to spend some time working more directly within their interest areas.

> **Hack days and 20% time**
>
> **Hack days** are miniature hackathons where participants spend a day focusing on solving a problem in a new way and bootstrapping a solution they can demo to the team at the end of the day. **20% time** is the practice of giving team members some percentage of their working schedule to work on relevant self-directed projects. These techniques are targeted applications of autonomy that can help increase engineer motivation and enthusiasm.

Particularly at senior levels, engineers are more likely to feel like there is no one they can learn from in their immediate vicinity. If you are working in a company with 50 engineers or more, you may be able to change this by introducing communities of practice. **Communities of practice** enrich working environments where engineers are organized around products or projects by creating professional structures for technology interest areas. For example, if your engineering team is organized by product lines into search, discovery, and content, it may be the case that each of those teams has one or two native iOS app developers. For those developers, you could establish a community of practice for iOS app development where engineers across teams gain access to rituals, resources, and mentorship opportunities organized around their specific interests. Communities of practice can increase satisfaction by improving access to growth and development resources in particular interest areas, such as programming languages and platforms.

Self-actualization

Many engineers seek fulfillment by striving to reach their full potential in their work. Set up time to connect with your team members to learn their technical and professional goals. Engineers on your team may be goal-oriented but not self-directed enough to establish and work toward those goals without your help and guidance.

If your workplace offers individual goal setting and growth milestones, take full advantage of that. In a busy workplace, it is easy to rush through goal setting, but it is much more impactful when you apply your full effort to the process. Don't just go through the motions; work with your team members to set goals thoughtfully and with intention. It is important to both listen to them and observe what they lean toward and demonstrate potential in. Your goal as an engineering manager is to help each member of your team reach their full potential, but this is often not a straightforward and obvious path. It involves missteps and failures and, sometimes, resistance and frustration. To maximize their satisfaction, help and support them in earnest throughout this journey.

Satisfaction with their manager

It is a commonly repeated adage that people don't leave jobs—they leave managers. This has been disputed in research (`https://www.cultureamp.com/blog/biggest-lie-people-quit-bosses`), but anecdotally, many engineers can name a time that they left a position primarily because they couldn't resolve differences with their direct manager. Satisfaction with manager behaviors is a contributing factor to overall job satisfaction.

Throughout this book, we have established a set of behaviors and practices for engineering managers that are satisfying and beneficial to engineering teams. Connecting with your team, demonstrating bravery and integrity, being accountable, setting expectations, communicating well, caring about your team, and facilitating positive team emergent states such as psychological safety and trust are all practices designed to produce satisfaction with managerial behavior. If you have followed the practices outlined in this book, you are already working to increase satisfaction with your management style. Build on this by seeking feedback and making your team feel supported and appreciated.

Feedback

Everyone is different and may respond to your actions differently, so it is important to get feedback on your management style and practices. As we mentioned in *Chapters 1* and *9*, engineering managers need to seek out feedback on what may be missing from their leadership behaviors in the eyes of their team members. This can be done by directly asking your team members in their one-on-ones, but it is also helpful to provide anonymous paths for your team to give feedback if they do not feel comfortable telling you directly. Anonymous paths can include satisfaction surveys or a digital suggestion box in a simple web form, such as `jotform.com`. You can also ask your manager to conduct **skip-level meetings** and report back any feedback they received.

Feeling supported

To increase satisfaction with you as a manager, be sure that your team members feel supported. There is an important distinction here between *being* supported and *feeling* supported. It may be the case that you are willing to support your team members however you can, but if they do not know that, it might leave something to be desired.

Everyone wants to feel that their manager is there for them when needed. We want our manager to have our back. We want to believe that our manager is our advocate, believes in us, and is willing to stand up for us and protect us if that is required. This is one of the reasons why it is so important to assume the best of our team members, as we learned in *Chapter 8*; when you make a wrong assumption, they may feel betrayed and abandoned. Feeling supported relies on mutual trust. To make your team members feel supported, continuously demonstrate caring, integrity, and trustworthiness.

Feeling appreciated

In addition to feeling supported, team members also want to feel appreciated by their manager. In other words, we want to feel that we are seen in the way that we see ourselves. We want our efforts to be recognized and valued. We want to be acknowledged as an individual and not treated as an interchangeable cog in a machine. We want to feel that we are seen for our own unique traits we bring to the table. To increase satisfaction with your leadership style, find ways to see, appreciate, and acknowledge your individual team members. For example, call attention to something unique they have contributed to their work and thank them for it. Make sure they know you appreciate them individually, not just as another engineer.

Beyond satisfaction with you as an engineering manager, do what you can to influence satisfaction with the company as a whole.

Satisfaction with the company leadership and direction

To some degree, engineers are attuned to overall company leadership. What a company does and how it does it can be a major factor in overall job satisfaction. This can vary strongly from person to person, but many people want to work in an environment where they have high confidence in the leadership and direction of the business.

As an engineering manager, you probably don't have direct control over your company's leadership and direction, but since they impact your team, these factors should still come under your consideration. You may be able to influence or advocate for practices that increase your team members' satisfaction with the company. Areas to consider include the following:

- **Values alignment**: Your company will either have formalized values or unspoken values that are communicated by the choices that company leaders make and what they put emphasis on. Assess how well these values align with those of your engineering team. See whether you are able to translate these values for your team and frame them in a way that makes them better appreciated. Talk to your manager and senior leaders in the company about areas where you think values could be improved.

- **Company purpose or mission**: Most companies have a clearly communicated mission or purpose, but some may feel like they juggle competing priorities. If you have a clear mission, help to translate it to your team to characterize it in aspirational terms. If you lack a clear mission, try to carve one out that is relevant to your team to give them a strong sense of purpose in their work.

- **Reputation**: You most likely have very little control over how your company is viewed publicly, but this can be a factor in company satisfaction. If you find you have a problem with the company's reputation, it is important to be prepared to have conversations about this and dispel any wrong notions.

- **Leadership behaviors**: How executives behave at your company can be a significant factor in company satisfaction. Look for ways to understand, contextualize, and translate the actions taken by your company's senior leadership in terms that make sense to your team. If you find yourself disagreeing with leadership practices, start by talking to your manager to better understand the dynamics. Look for ways to advocate for leadership behaviors that are beneficial to your team.

Raise your awareness of your team's satisfaction level with the company and do what you can to increase it.

Now that you have a sense of how to retain talent by increasing satisfaction for your team, let's learn how to apply these techniques judiciously.

Pitfalls of retaining talent

Retaining talent on your team is not universally desirable. While it is generally a good thing to do, there are limitations and drawbacks to keep in mind. In your efforts to retain talent on your team, take care to avoid scenarios where turnover is too low or you put too much emphasis on engineer satisfaction.

Can turnover be too low?

When the engineers on your team are happy and productive, it is a good thing. You want your engineers to grow and develop over time and make great contributions to your code bases and products. Part of growing and developing is eventually growing into new responsibilities and roles. Your team members might do so within your company or they might not, but either way, growth is important. Most (but not all) of your engineers should progress from their roles eventually. As an engineering manager, it is your goal for them to do so in a gradual way such that you hold on to the same levels of knowledge and expertise within the team. This gradual change over time allows the team to benefit from fresh perspectives and ideas.

Zero turnover on your team can be dangerous when team members become too comfortable or stuck in their ways. It is hard to think about problems in new ways when you are so used to solving them in the way you have in the past and are too comfortable in your routines. Having some fresh perspectives on a team helps to keep everyone on their toes.

Think about retaining talent such that your team members eventually graduate from their position rather than becoming frustrated or disengaged and departing suddenly. Graduating from your team should be a slow process, where they have ample time to inform newer team members and move on amicably. This is the eventuality of self-actualization for the majority of teams and positions. Your team composition should follow a slow cycle of new talent, growth, service, and eventual progression beyond their role.

Can engineers be too satisfied?

In *Chapter 2*, we introduced the servant leadership style and noted that one of the drawbacks of this style is that it is vulnerable to subjective expectations of team members. This is to say, while it is usually a good thing to serve your team and work toward their satisfaction, you must first agree that their expectations are reasonable and appropriate.

When teams are catered to, a few team members might eventually develop overgrown expectations. As individuals' needs are met, a few of them might continue to develop new needs that evolve to the point of absurdity. It is one thing to be satisfied and appreciated but another to feel entitled. Engineering managers must pay attention to the effects and behaviors that result from their efforts to give their teams job satisfaction. If the cost of satisfaction moves beyond reasonable accommodations into the realm of frivolousness, it's time to pull back.

Unreasonable expectations need to be addressed immediately and directly. Use radical candor to tell your team member your point of view directly in their one-on-one. Always give them the benefit of the doubt and ask questions to avoid jumping to conclusions. If a team member expects something well outside of the realm of possibility—for example, being sent to four conferences in a year—you might have a conversation by asking, "*How did you come to expect to be sent to so many conferences this year? I want you to feel like you have learning opportunities, but to be completely honest what you are asking for seems unreasonable to me.*" Refer back to *Chapter 8* for a refresher on setting expectations.

Keep your team satisfied while also providing awareness of the effort that goes into providing for them. When they know that you work hard to support them, they will appreciate you as much as you appreciate them.

Summary

In this chapter, you learned how to focus on the long-term success of your team by retaining talent. In the competitive landscape of software development, retaining the talent on your team can give you an edge. Here is a recap of what we covered in this chapter:

- The turnover rate refers to the rate at which positions are vacated and rehired

- Talent retention is a useful skill to keep costs lower, keep productivity higher, retain organizational know-how, and be more competitive in hiring

- To retain talent, first use the techniques outlined throughout this book to create a desirable work experience, and then focus on increasing the satisfaction of your team members:

  - Increase satisfaction with the workplace environment by providing a work experience that incorporates Maslow's hierarchy of needs. Meet both the collective and individual needs of your team members.

  - Increase satisfaction with growth and opportunities by paying special attention to cycles of esteem, access to learning resources for cognitive needs, and goal setting for self-actualization.

  - Increase satisfaction with you as an engineering manager by collecting feedback and making your team members feel supported and appreciated.

  - To increase satisfaction with a company and direction, look to values, mission, reputation, and leadership behaviors.

- The pitfalls of talent retention include the turnover rate being too low and team expectations becoming excessively high:

  - Encourage a gradual cycle of growth and promotion beyond your team to avoid over-retaining and becoming stagnant.

  - As you raise satisfaction in your team, be careful to avoid excessive expectations. Address unreasonable expectations immediately.

Now that you have a foundational understanding of how to retain talent, we will close out the book by providing resources for further exploration of long-term strategies in *Chapter 16*.

Further reading

- *To have and to hold* (`https://www.shrm.org/hr-today/news/all-things-work/pages/to-have-and-to-hold.aspx`)

- *It costs $50k to hire a software engineer* (`https://cgroom.medium.com/it-costs-50k-to-hire-a-software-engineer-d06a0d051abf`)

- *The cost of replacing an employee — it's more than you think* (`https://resources.workable.com/stories-and-insights/the-cost-of-replacing-an-employee`)

- *The 4 "perks" good developers really want* (`https://venturebeat.com/business/the-4-perks-good-developers-really-want/`)

- *It's time to reimagine employee retention* (`https://hbr.org/2022/07/its-time-to-reimagine-employee-retention`)

- *Six tips for building communities of practice* (`https://www.software.ac.uk/blog/2021-08-03-six-tips-building-communities-practice`)

- *A pragmatic guide to starting communities of practice in the organization* (`https://dev.to/optiklab/a-pragmatic-guide-to-starting-communities-of-practice-in-the-organization-25n3`)

16

Team Design and More

Over time, engineering managers have opportunities to reshape their engineering teams, reconsidering how they are organized and oriented to their work. This reshaping might be gradual, as managers work to guide engineers and influence their team climate and emergent states, or it may be sudden in response to organizational change.

Engineering team design refers to how you structure your engineering team, its roles, and how those roles operate. Team design optimizes productivity and other success metrics, such as efficiency, effectiveness, and innovation. You might have an opportunity to design your team from scratch during an organizational change event, but more often, team design involves incremental changes to solve team problems that crop up in day-to-day work.

In this chapter, you will learn the foundational concepts in designing engineering teams. You will learn about team structures and characteristics and revisit Conway's Law. You will receive answers to some lingering questions that have not yet been broached. By the end of this chapter, you will understand the most common approaches to team design and have a list of resources to further support you once you have finished reading this book.

This chapter is organized into the following sections:

- Engineering team design
- Lingering questions

Let's start by gaining an understanding of how we might design our engineering team.

Introducing engineering team design

New engineering managers rarely have the opportunity to directly choose a team design, but regardless of where you are in your manager journey, having awareness of team design concepts can help to orient you and reveal solutions to problems. Depending on your work context and goals, team design can make a big difference in your team's productivity and performance. To that end, let's start by learning some of the most common software development team designs and how they operate.

Common team structures

Software development organizations structure their teams in different ways to meet organizational needs. In terms of reporting hierarchies, these are the most common approaches for team structure:

- **Functionally aligned teams** are grouped by a particular skill, which may be a technology or type of work, such as an *iOS development team* or an *infrastructure engineering team*

- **Product aligned teams** have members who are dedicated to a product area and, regardless of their functional expertise, report to a manager for that product, such as a *search product team*

- **Matrixed teams** are a hybrid of functional and product alignment, where the engineers often report to a functional manager but work day to day in a cross-functional product team

In the next subsections, we will explore these team structures in more detail.

Functionally aligned teams

In functionally aligned teams, you typically see a team built around a particular skill area with an engineering manager who is knowledgeable in that skill, as shown in the example in *Figure 16.1*.

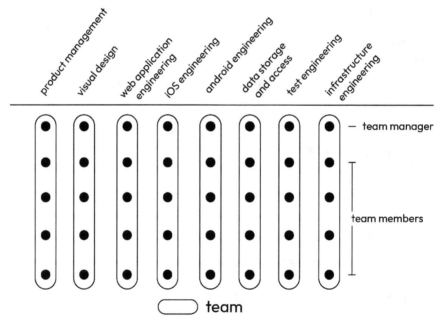

Figure 16.1: An example diagram of functionally aligned teams

In a functionally aligned team structure, the team takes on work that fits within their skill area regardless of the product or project. Team members work across many simultaneous projects to contribute to the portions that match their team function.

Functionally aligned teams make it easier to keep conventions and practices consistent across many projects and products. Since a team is built around a skill area, bonds and allegiances are primarily to that function, encouraging a depth of mastery, mentorship, and excellence in that function. The engineering manager can focus on developing and formalizing the practices and processes for their function, streamlining the team's ability to integrate with and provide for projects. Engineers on these teams often find it preferable to receive performance evaluations from a leader who can understand their subject matter expertise and the depth of their contributions.

Conversely, it can be more challenging with functionally aligned teams to make work meaningful, since it typically involves hopping in and out of many projects. Engineering managers often struggle to understand the nuances of project and product needs when there is a wide range of simultaneous efforts. Functionally aligned teams can also suffer from emphasizing the function rather than the final product. It can be a double-edged sword to improve functional quality at the cost of product quality, so managers must work hard to ensure this isn't the case.

From a management perspective, the biggest problem with functionally aligned teams is that they can completely lose their team identity. A team is a group of people working together toward a common goal. Functional teams may not work cohesively enough to feel like they share the same goal. Without a team identity, you don't have the foundation to work towards positive team emergent states and a productive team climate. For functionally aligned teams, engineering managers can reinforce team identity by doing the following:

- Giving the team an overarching meaning and purpose
- Establishing team values
- Holding regular team rituals
- Facilitating relationships and bonding
- Providing collaborative tasks or pairing programming sessions

Product aligned teams

Product aligned teams are built around products rather than skills, so there can be considerable variation in their makeup and leadership. Usually, product teams are led by a manager who is knowledgeable in the product area, often a product manager or product engineer. As shown in *Figure 16.2*, the team consists of members with the right functions to deliver the product, including specializations within engineering, design, data, infrastructure, and many other areas. Team members are dedicated to working through a roadmap for their product area only.

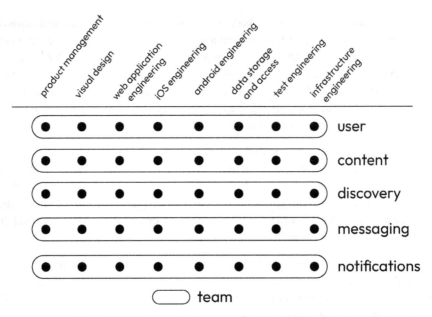

Figure 16.2: An example diagram of product aligned teams

Product aligned teams benefit from making the end product their focus and first-class citizen. Engineers on these teams develop deep product expertise and specialized skills in a product area. For example, if your team focuses on a subscription product, they will develop skills in growth engineering (growing subscription rates). Over time, these engineers iterate on the product to produce more sophisticated solutions and approaches.

On the downside, it can be harder to keep engineering conventions and practices consistent across product aligned teams. Engineers might work less efficiently because they are unaware of what they might be able to reuse from similar projects. Engineering quality can be harder to maintain when the focus is on the product, since team structure and expectations are built around product needs rather than engineering needs. For example, an engineer might be consistently praised by a non-technical product team leader for their work, while delivering subpar code quality. It can also be harder to receive guidance and mentorship in a particular functional area when functional peers are not immediately available. In product aligned teams, establishing functional communities of practice can help to alleviate many of these drawbacks. Alternatively, matrixed teams are another solution for the same purpose.

If you are the manager of a product aligned team, then you manage team members from multiple functional areas. It can be difficult to understand and effectively guide such a wide range of contributors. To do their best work, team members need to feel equally valued and respected. Support cross-functional team members by doing the following:

- Defer to your subject matter experts. Take care to avoid making assumptions or underestimating effort. Don't let your ego get in the way.

- Develop a support network of cross-functional partners. Have colleagues to go to when you need feedback, a gut check, or a sounding board.

- Develop t-shaped skills. While you can't be an expert in everything, it helps to understand your team members' functional areas as well as you can. (`https://wordspy.com/words/t-shaped/`).

Matrixed teams

Matrixed teams incorporate both functional and product structures. Often, each engineer reports to an engineering manager of the same function and separately has a "dotted-line" reporting or is simply assigned to work with the head of a product area, as shown in *Figure 16.3*:

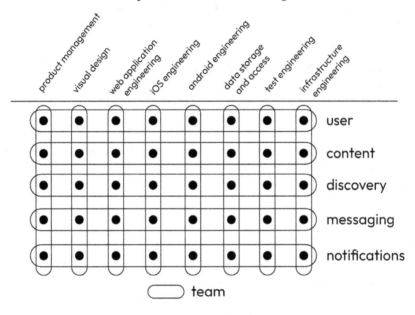

Figure 16.3: An example diagram of matrixed teams

Matrixed team implementations strive to deliver benefits and lessen the drawbacks of product aligned and functionally aligned teams. They contain a functional hierarchy to encourage skill-based alignment, paired with a product hierarchy to produce iterative product learning and product focus.

Matrixed teams introduce some of their own drawbacks, since this structure is more complicated to manage. They require more coordination between leaders to make sure that direction is aligned. At times, managers from engineering and product may fall out of alignment in their efforts and end up competing with one another, leaving engineers on the team confused and feeling like they have too many bosses.

Matrixed organizations can feel noisy and chaotic to team members who are on the receiving end of expectations arriving simultaneously from multiple directions. At its worst, this sort of environment is associated with reduced engagement, which, in turn, affects product outcomes negatively. To be successful, matrixed teams need the following:

- Aligned leaders to provide clarity
- Unambiguous roles and ownership
- Open lines of communication
- Well-defined processes

Matrixed teams require a more conscious effort to ensure that operations are clear and straightforward so that engineers aren't overwhelmed by meetings and updates. Open communication between leaders helps to quickly resolve disagreements and keep teams moving forward.

Now that you understand the common team design structures, let's look at how personal characteristics can help with team design.

Team characteristics

In addition to functional skills and hierarchical structures, engineering team design can benefit from consideration of team member characteristics, such as organizational tenure and personality type.

Organizational tenure

Organizational tenure is the amount of time a team member has spent at an organization. As we mentioned in *Chapter 15*, tenure matters because organizations have learning curves such that the newest team members might struggle to achieve their objectives. Learning curves are steeper when organizations are larger and when they have more complex structures, such as matrixed teams. Ideally, teams should have a mix of members with different tenures and include some with longer terms to provide additional context and organizational know-how to the team. New team members bring new perspectives, ideas, and experiences to draw from.

Personality types

Personality is another important consideration in team design. In addition to their skills, individuals possess personality traits that affect how they work on a team. In different team contexts, individual traits can make a major difference in team outcomes.

In his research on workplace teams, Dr Meredith Belbin developed a theory on the personality roles that team members demonstrate called **Team Role theory**. As shown in *Figure 16.4*, this theory asserts that team members fall into three personality categories—action-oriented, thought-oriented, and people-oriented.

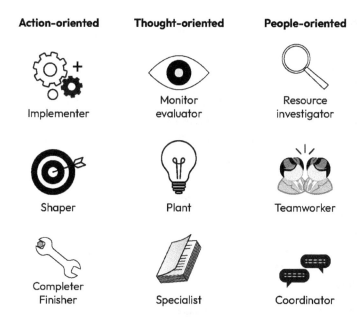

Figure 16.4: Belbin's team roles

Belbin described these categories as follows:

- **Action-oriented roles** keep things moving and tend to do better with deadlines. These roles include *shapers* who drive the team forward, *implementers* who are prolific in their contributions, and *completer finishers* who strive for perfection.

- **Thought-oriented roles** continuously generate new ideas for improvement. These roles include *monitor evaluators* who are rational, systematic thinkers, *plants* who are innovative creative thinkers, and *specialists* who are single-minded deep thinkers.

- **People-oriented roles** are strong communicators who help maintain a team's social fabric. These roles include *resource investigators* who seek out and build bridges with outside communities, *teamworkers* who are extroverts and natural collaborators, and *coordinators* who use communication skills to build consensus and alignment.

Research indicates that teams that include a mix of these roles produce more successful outcomes compared to those that have a single type. Different teams are best served by different combinations of team roles to suit their objectives. As you consider the design of your team, consider which of these roles might be the most impactful to the needs and aspirations of the team. For example, in a complex matrixed team in a large organization, having a *resource investigator* personality on the team might help the entire team to stay connected to the organization and navigate cross-team challenges.

In addition to these team member characteristics, team design also has a relationship with software design.

Conway's law—part 2

In *Chapter 4*, we introduced Conway's Law:

"Any organization that designs a system (defined broadly) will produce a design whose structure is a copy of the organization's communication structure."

As you learned in *Chapter 4*, you can mitigate the effects of Conway's Law by strengthening communication across teams, incorporating modular design, and maintaining consistent interfaces.

While you have the ability to implement strategies to limit the unwanted effects of Conway's Law, it is worth noting that there is a natural tendency for the boundaries of systems to line up with the boundaries of teams. Given that this is the case, it follows that when you design teams, it can make sense to structure them in a way that directly conforms to your current or desired system design. This has been dubbed the **inverse Conway maneuver** (`https://www.thoughtworks.com/en-us/insights/podcasts/technology-podcasts/reckoning-with-the-force-conways-law`). Designing teams to mirror a desired architecture is often seen in product aligned team structures.

Now that you understand the foundational aspects of team design, we will wrap up by answering some lingering questions you might have.

Lingering questions

In this handbook, there is a broad range of information to help engineering managers do their best work, but there are a few topics not covered in previous chapters. Along with team design questions, you may have a few more lingering questions, which will be addressed here.

What are squads, chapters, guilds, and tribes?

If you have heard these terms used in the context of team design, you might have heard them in reference to **the Spotify model**. In the early 2010s, Spotify wrote extensively about its team design approach and popularized a structure consisting of a *product aligned* organization with formalized *communities of practice*, as shown in *Figure 16.5*:

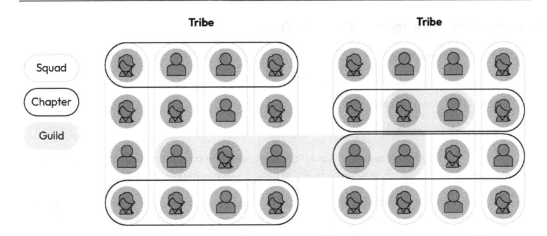

Figure 16.5: The original Spotify model of team design

In the Spotify model, a *squad* is a cross-functional team working together on a specific product area. A *chapter* is a small community of practice with a narrow area of focus. A *guild* is a large community of practice that works across a larger portion of the organization. *Tribes* are groups of squads working on related features or goals. Chapters work within a single tribe, while guilds work across multiple tribes.

Many companies use an adapted version of the Spotify model, or simply use terms such as *squad* and *guild* in their own context. Since each company takes its own approach to team design language, these terms do not generalize well across companies, and you typically need to ask companies to describe their usage of organizational terminology.

How many engineers can one person effectively manage?

The size of your engineering team is often a function of budgets and the necessary delivery capability, but as your team grows to meet those needs, you may wonder what a reasonable size for one person is. **Span of control** is a term that refers to how many direct reports a manager is responsible for. Historically, the figures shown in research to be optimal for efficiency and effectiveness are around seven direct reports (DOI: `10.1108/02756660310494854`). Beyond seven, the quality of management has been found to degrade.

More recently, there has been attention given to the effects on span of control by management style. A coaching management style is more time-consuming and necessitates a smaller team of three to five. A delegating management style can handle a larger team of 8 to 10. The size of the team can be nudged up or down, based on the amount of guidance given to the team members.

Other factors identified in research as impacting the span of control include the following:

- How complex the work is
- How repetitive or standardized the work is
- The level of coordination required
- Task interdependence

For example, if you manage a team that sets up the same software platform for multiple clients, you could more easily manage a larger team than if you built custom software for each client.

Determine your maximum effective span of control by considering the demands of your management style, the complexity and repetitiveness of your work, and how much time you spend on coordination. Start with seven, and add or subtract one for each of those concerns.

What long-term goals should I have for my team?

With all of the demands on your time as an engineering manager, it can be difficult to find the space to think about what you would like to accomplish holistically for your team in the long term. For most teams, a good starting point is to set the goal of making your role as engineering manager superfluous. In other words, make your team capable of *self-sufficiency*, where peers exhibit the leadership and mutual support to deliver optimal outcomes and resolve problems with no intervention from you whatsoever. This doesn't mean ignoring your team or losing your connection with them but, instead, giving them the autonomy and confidence to solve their own problems. This is good for your team because it gets them operating at their highest level, maximizing their growth. This is good for you because it frees you up to reorient your role toward strategic activities, such as research, developing a strategic partnership, or theorizing a new prototype.

Realistically, your team will always need you some of the time, but striving to make your involvement obsolete helps you to shape your team into a well-oiled machine and produces a legacy that you can be proud of.

Another key goal for teams is *impact*. Impact is a holistic approach to assessing outcomes—what the team achieved in relation to the expectations. Is the team simply meeting basic expectations, or are they making a difference that goes beyond requirements? Impactful teams have a positive effect that extends beyond the bounds of their ownership domains. They contribute to and advance the organizational vision with thought leadership and careful decisions.

Because it is subjective and context-dependent, impact is a complex goal that incorporates qualitative measures and individual judgment. A good way to consider impact is in relation to quarterly goal-setting, such as OKRs.

What exactly is engineering culture?

Culture and **culture fit** are terms you hear often in relation to recruiting and hiring. The meaning of these terms is fairly ambiguous and varies in real-world usage. They can be convenient terms summarizing your engineering values, principles, team climate, and general atmosphere. Conversely, they are sometimes used as hand-wavy explanations or excuses as to why a course of action is being taken, as in "*that candidate wasn't a good culture fit.*"

While *culture* can be a convenient umbrella term to encompass the components of your engineering ethos, it is a good idea to pay attention to its usage to avoid glossing over important details. Avoid saying someone isn't a good culture fit for the team (unless you are intentionally using a euphemism) because it lacks specificity and can imply bias. If an interviewer on your hiring team tells you a candidate isn't a culture fit, the immediate question is, "*Why?*" It is better to be clear by saying, "*I don't think the candidate matched well with these three values because…*"

What should I do if I disagree with my manager?

It is best to find a position where you can have a good working relationship with your manager. An ideal position is one where you and your manager are in agreement on working practices and you have an opportunity to learn from them. You don't need to be friends with your manager, but mutual respect is necessary for a positive and productive working relationship.

When you disagree with your manager on any situation or a course of action, you can first treat it like any other stakeholder disagreement, seeking to understand where their view comes from. Why does your manager hold that particular opinion? What is it based on? Probe the situation with earnest curiosity and zero frustration or antagonism. For example, you might say any of the following:

- *I'm trying to wrap my brain around this new direction so I can get my team going. Can you help me understand how you came to this decision?*

- *I'm not pushing back here, but it would help me to understand what you see as the benefits of doing it this way.*

- *It would really help me in my growth as a manager to see this more from your point of view. What is your thought process on this?*

When you seek to understand your manager's point of view, how you deliver your questions can make all the difference, so be sure to do so when you are in the right frame of mind to be open and free from emotion. After this conversation, you have three possible outcomes:

- Your manager explained their point of view and you now agree with them

- Your manager explained their point of view and you still disagree with them

- Your manager told you to kick rocks and refused to explain anything

If you find that you still disagree with your manager, the next step is to see if you can change their mind. Every manager is different, but you may want to assume that you will only get one or maybe two chances to change their mind, since that is often the case. For this reason, it behooves you to be well prepared for the conversation. Collect the best data you can find to provide proof that your point of view is not just an opinion but also substantiated. With that data, make your best, clearest argument for why specifically you disagree with the explanation they gave you and why your approach is preferable. Be earnest and succinct but not pushy or whiny. For example, you might say, "*I know you think that _____, but I think it's worth considering _____ instead because…*" or, if you have a manager that is prone to interrupting, "*I know your view is that we should _____, but I want to ask you to give me five minutes to explain why I think we should consider _____.*"

If your manager still disagrees with you or you never made it past the kick rocks stage, then you must accept that this one didn't go your way. **Disagree and commit** to the choice that has been made. As *Ask a Manager's* own Alison Green once said, "*This is the nature of having a manager.*"

Summary

In this chapter, you learned the foundations of engineering team design and how to adjust your team to suit your needs:

- *Engineering team design* refers to how you structure your engineering team, its roles, and how those roles operate
- The three primary engineering team structures are functionally aligned, product aligned, and matrixed:
  - Functionally aligned teams are built around a particular skill area with an engineering manager who is knowledgeable in that skill
  - Product aligned teams are organized according to product areas rather than functional skills
  - Matrixed teams incorporate both functional and product structures
- Communities of practice help to maintain skill-based conventions and communication in product aligned organizations
- Team characteristics to consider in team design include organizational tenure and personality traits:
  - Teams with a mix of different tenures help each other to learn about the organization and new approaches
  - Belbin's personality categories are *action-oriented*, *thought-oriented*, and *people-oriented*; a mix of these personalities can better support each other and produce better outcomes
- Team design can also benefit from considering software design and interfaces, per Conway's Law

- Squads, chapters, guilds, and tribes often refer to the Spotify model but cannot be assumed to have a consistent meaning

- Span of control is a manager's number of direct reports; a manager's ideal span of control should be five to seven and adjusted up or down, based on the amount of guidance needed, complexity, repetitiveness, and additional needs

- Good long-term goals for all engineering teams are self-sufficiency and impact

- *Engineering culture* refers to values, principles, team climate, and atmosphere, but the term is often overused or misused

- When you disagree with your manager, first seek to understand and then to be understood (read more at `https://www.franklincovey.com/habit-5/`)

- If you can't gain support for your point of view, *disagree and commit* to the decision made

With that, you have finished the *Engineering Manager's Handbook*. I hope these techniques serve you well throughout your career.

Further reading

- *"Understanding team design characteristics"* from *Organizational Behavior*, 2010 (`https://open.lib.umn.edu/organizationalbehavior/`)

- *Belbin's team roles* (`https://www.belbin.com/about/belbin-team-roles`)

- *How to Build Your Own "Spotify Model"* (`https://medium.com/the-ready/how-to-build-your-own-spotify-model-dce98025d32f`)

- *Davison, Barbara. "Management span of control: how wide is too wide?."* Journal of Business Strategy 24.4 (2003): 22–29 (`https://doi.org/10.1108/02756660310494854`)

- *Reckoning with the force of Conway's Law* (`https://www.thoughtworks.com/en-us/insights/podcasts/technology-podcasts/reckoning-with-the-force-conways-law`)

- *Measures of engineering impact* (`https://lethain.com/measures-of-engineering-impact/`)

- *Disagree and commit: the importance of disagreement in decision making* (`https://hackernoon.com/disagree-and-commit-the-importance-of-disagreement-in-decision-making-b31d1b5f1bdc`)

- *On disagreeing with your manager* (`https://www.askamanager.org/2009/03/when-you-disagree-with-your-boss.html`)

Index

A

A/B testing 209
accountability 151
  big mistakes, handling 151
  mistakes by engineers, handling 152
  successes, handling 152, 153
Accountable, Responsible, Participant,
  Advisor (ARPA) 70
accountable team culture
  accountability 151
  building 145
  guidance, providing 145
  ideal behaviors, providing 148
action-oriented roles 235
assessing teams, pitfalls 127
  Goodhart's law 127
  McNamara fallacy 128
  perverse incentive 127
awareness, raising against reliability 85
  metrics 86
  metrics, active communication 86
  metrics, passive communication 87
AWS Fault Injection Simulator
  (AWS FIS) 159

C

candidate assessment 193
  feedback, measuring 193, 194
candor breaks 178
  reference link 178
CAR format 122, 123
championship team 187
churn 207
commitment, building to reliability 82, 83
  influencing 85
  ownership mindset 83, 84
  sense of pride 84
communication 109
  principles 110-116
communication structure 116
  depth 117
  duration 117
  format 116, 117
  urgency 118
communication, with engineering team 118
  group communications 120
  one-on-one meetings 119, 120
  personal commitments 121

communication, with leadership team 121

broader value, conveying 122

storytelling 121, 122

uncertainty, reducing 123

company leadership and direction satisfaction 225

Conway's Law 55, 236

cross-functional leadership

demonstrating 98

cross-functional partners 98-190

aligned structure, providing 101

aligning with 100

roles, providing 101

same-team attitude, adopting 100

team visions, uniting 100

cross-functional partners, situations 104

escalating 106

issues, challenging 104

issues, learning 105

working 105, 106

cross-functional teams 97

culture fit 239

D

DevOps Research and Assessment's (DORA's) 128

measures 129

dot voting 205, 206

E

Eisenhower method 204, 205

engineering culture 239

engineering leadership style 17

engineering leadership style archetypes 21

coach and delegator 22, 23

commander and servant 21, 22

communicator and co-creator 23-25

engineering manager 186, 229

activities 7, 8

day-to-day life 8-13

long-term goals for team 238

preparing, for responsibilities of 14, 15

responsibilities 4-6

engineering manager (EM), architecture concern 47

factors 47, 48

naming 51, 52

open source 51

ownership and maintenance 49

engineering manager (EM), architecture process

affect heuristic 54

availability heuristic 54

building 53

emotion and biases 53

managing 52

positive hypothesis testing 54

representativeness heuristic 54

stages 53

engineering manager (EM), architecture setting 44

building blocks 45

decision methodology 46

environment 44, 45

information 46

engineering manager (EM), ownership and maintenance

trade-offs 49, 50

engineering managers, failure modes scenarios

decision making 35

drawbacks 36

information sharing, with team 33, 34

leadership 30, 31

management 32, 33
    role and responsibilities 37, 38
engineering team design 229
    functionally aligned teams 230, 231
    matrixed teams 230, 233, 234
    organizational tenure 234
    personality 234
    product aligned teams 230-233
    size 237
    team characteristics 234
engineering teams
    assessing 126, 131-133
    qualitative measures 129, 130
    quantitative measures 128, 129
    scaling 185
    success definition 130, 131
engineer partners 189
estimations 63
express, demonstrate, incorporate,
    and pay tribute (EDIT) 176

F

fragile team 170
functionally aligned teams 230, 231

G

Goodhart's law 127
growth and opportunities satisfaction 221
    cognitive 222, 223
    esteem 221
    self-actualization 223
guidance, providing
    access to information, providing 147
    clear goals, maintaining with
        success definitions 146
    resources to information, providing 147

    roles and responsibilities, establishing 147
    shared understanding of work,
        building 145, 146

H

hiring 186
hire process
    handling 198
hiring team
    building 188
    cross-functional partners 190
    engineer partners 189
    recruiting partners 188, 189
hundred-dollar test 205

I

ideal behaviors, providing 148
    accountability, creating for 149, 150
    accountability, demonstrating 148
    authority, sharing 150
incident reports 93
interviewing practices 190
    interview process setup 190, 191
    non-technical interviewing 192, 193
    technical interviewing 191, 192
interview process
    setting up 190, 191
interview round 191
inverse Conway maneuver 236

J

job profile
    example 187
    writing 187, 188
just-in-time documentation 196

K

kanban 208

L

language 109
large team management 198, 199
    delegation 199
    focus on people, over projects 200
    long-term plan 199
    prepare to switch contexts 200
    short-term plan 199
    team members, accessing 200
leadership style 17, 25
leadership style, origins 18
    management theory origin 20, 21
    natural origin 19
    philosophical origin 19, 20

M

manager hand-off meetings 212
managing risks 155
Maslow's hierarchy 220, 221
matrixed teams 230, 233, 234
McNamara fallacy 128
metrics 86
minimum viable product (MVP) 209
MoSCoW prioritization 205

N

Net Promoter Score (NPS) 129
non-technical interviewing 192, 193
numerical prioritization 205

O

onboarding 195
    automation, using 195, 196
    structured performance objectives,
        providing 196, 197
    new-hire training, providing 196
on-call rotations 91
Open Worldwide Application Security
    Project (OWASP) 163
organizational changes 210
    alliances, building 212
    directive 211
    managing 210
    momentum, building 212
    stakeholders, aligning 212
    team 210
    team, aligning 211
    team guidance 211
    team support 211
organizational tenure 234

P

Parkinson's law 63
people-oriented roles 235
perverse incentive 127
pivot 210
planned service interruptions 91, 92
planning and delivery stages
    environment 61
    goal orientation 62
    setting 60
power posing 174
pre-mortem 67
principles of communication 110-116
prioritization 63, 204

prioritization methods 204
  dot voting 205, 206
  Eisenhower method 204, 205
  numerical assignment 205
  selecting 206
  stack ranking 205
priority changes
  managing 206
  prioritization context 207
  prioritization dynamics 207
  prioritization solutions 208, 209
  user testing 209
priority churn
  effects 208, 209
  team impact 208
product aligned teams 230-233
production systems 81
project delivery 71
  friction, removing 73
  problems and solutions 76
  project kick-off 71
  project scope 77
  user stories 72
project delivery, friction
  people friction, removing 74, 75
  process friction, removing 73
  removing 73
  tooling friction, removing 75
project planning 63
  estimation 63
  need for 60
  plan, forming 69
  prioritization 66
  risks, assessing 67
  roadmapping 68
project planning, estimation
  approaches 63, 64
  assumptions 65, 66

  work, breaking down 64, 65
project planning, plan form
  change readiness 69
  communication strategy 70
  project contributor roles 70
  stakeholder clarity 69
project planning, prioritization 66
project planning, risks assessment 67
  risk communication 67, 68
  risk identification 67
pull request (PR) 129, 195

Q

qualitative data 128
qualitative data sources
  consideration 129
quantitative data 128

R

RACI chart 70
radical candor 115
read-eval-print-loop (REPL) 191
**Recommendation, Establish Agreement,
    Perform and Execute, Provide
    Input, Decide (RAPID) 70**
recruiting 186
  mindset 186, 187
recruiting partners 188, 189
relationship building 102
  cross-functional knowledge 102
  feedback, providing 103, 104
  feedback, seeking 103, 104
  partners support 103
  time management 102

reliability
    awareness, raising 85
    commitment, creating 82, 83
    ownership mindset 83
reliability solutions 87
    alerting 91
    monitoring 90
    service interruptions 91
    service objectives 87
    support documentation 88
remote teams
    improving 141
reorg 210
resilient teams 170
    creating, with engineering
        manager's role 171
    need for 170
Responsible, Accountable, Supportive, 70
risk assessment
    completing 161, 162
risk identification 157
    chaos engineering 159
    common examples 157
    methods 159
    pre-mortem exercise method 159
    sources of risks 158, 159
    trust but verify method 159
risk prioritization
    risk likelihood 160
    risk mitigation effort 161
    risk severity 160
    risk tolerance 160
risk response plan
    validating, with 4 As 164
risk response, planning
    with 4Ts 163
    with Terminate 163
    with Tolerate 163

with Transfer 163
with Treat 163
risk response plan, validation
    Achievable option 164
    Actionable option 164
    Agreed option 164
    Appropriate option 164
risks 63
    managing 156, 164, 165
    responding 163
risks, managing 156
    risk identification 157
    risk prioritization 159
risk sources
    cognitive risks 159
    economic risks 158
    legal risks 159
    operational risks 158
    physical risks 158
    political risks 158
    social risks 158
roadmapping 63

S

satisfaction, with manager 223, 224
scaling
    engineering teams 185
scope creep 77
self-assessment 132
service interruptions 91
    planned service interruptions 91, 92
    unplanned service interruptions 92, 93
service-level agreements (SLAs) 87
service level indicators 88
service-level objectives (SLOs) 87
single point of failure (SPOF) 6, 157
site reliability engineering (SRE) 50

skip-level meetings 224
software architecture 43
span of control 237
Spotify model 236
  chapter 237
  guild 237
  squad 237
  tribes 237
stack ranking 205
stakeholder assessment 132
status pages 93
storytelling 121
subject-matter experts (SMEs) 186
sunk costs fallacy 172
support documentation,
    reliability solutions 88
  escalation paths 89, 90
  historical information and context 88
  known failure modes 89
  severity levels and expectations 89
  support goals or SLAs 89
  systems owners and contact information 89
  troubleshooting 89
systems designs 43

T

talent retention 218
  benefits 218
  company leadership
    and direction satisfaction 225
  growth and opportunities satisfaction 221
  pitfalls 226, 227
  satisfaction with manager 223, 224
  workplace environment satisfaction 219-221
team
  changes, preparing for 180
  classic stages 126

  communication, changing 180, 181
  leadership, changing 181
  marketing 194, 195
team dynamics 126
team emergent states 134, 135
  antecedents, for list of 136
  contextual factors 135
  fostering 136, 137
  performance and outcomes 134
  prioritizing 135, 136
team performance
  autonomy 138
  improving 137
  individuals, mentoring
    and coaching on team 139, 140
  mastery 138
  motivating 137
  purpose 138, 139
  remote teams, improving 141
team, preparing for change 171
  best practices, for building
    resilient culture 175-177
  best practices, for building
    resilient habits 178-180
  best practices, for managing self 171-174
team reorganizations (reorgs) 203
Team Role theory 234
technical interviewing 191
  code reading 192
  live coding 191
  systems design 192
  take-home tests 191
thought-oriented roles 235
timeboxing 60
top ten prioritization 205
T-shaped
  reference link 233
turnover rate 218

U

unplanned service interruptions 92, 93
user stories 72
    acceptance criteria 72
    characteristics 72
    framing information 73
    size 72
user testing 209

V

verbal communication 117
version control systems (VCS) 59

W

workplace environment satisfaction 219-221
written communication 117

Packtpub.com

Subscribe to our online digital library for full access to over 7,000 books and videos, as well as industry leading tools to help you plan your personal development and advance your career. For more information, please visit our website.

Why subscribe?

- Spend less time learning and more time coding with practical eBooks and Videos from over 4,000 industry professionals

- Improve your learning with Skill Plans built especially for you

- Get a free eBook or video every month

- Fully searchable for easy access to vital information

- Copy and paste, print, and bookmark content

Did you know that Packt offers eBook versions of every book published, with PDF and ePub files available? You can upgrade to the eBook version at packt.com and as a print book customer, you are entitled to a discount on the eBook copy. Get in touch with us at customercare@packtpub.com for more details.

At www.packt.com, you can also read a collection of free technical articles, sign up for a range of free newsletters, and receive exclusive discounts and offers on Packt books and eBooks.

Other Books You May Enjoy

If you enjoyed this book, you may be interested in these other books by Packt:

Squarespace from Signup to Launch

Kelsey Gilbert Kreiling

ISBN: 978-1-80181-308-2

- Build a website on Squarespace, step by step, with expert insights and practical tips

- Plan your site content with an easy-to-understand outline

- Source and create the visual elements necessary to achieve a professional website

- Go beyond pre-set templates by creating a polished design from navigation to footer

- Integrate custom code to enhance both the design and functionality of your project

- Optimize your website for mobile viewing and search engine visibility

- Implement effective marketing strategies to promote your site and grow your audience after its launch

Web Development with Julia and Genie

Ivo Balbaert, Adrian Salceanu

ISBN: 978-1-80181-113-2

- Understand how to make a web server with HTTP.jl and work with JSON data over the web
- Discover how to build a static website with the Franklin framework
- Explore Julia web development frameworks and work with them
- Uncover the Julia infrastructure for development, testing, package management, and deployment
- Develop an MVC web app with the Genie framework
- Understand how to add a REST API to a web app
- Create an interactive data dashboard with charts and filters
- Test, document, and deploy maintainable web applications using Julia

Packt is searching for authors like you

If you're interested in becoming an author for Packt, please visit `authors.packtpub.com` and apply today. We have worked with thousands of developers and tech professionals, just like you, to help them share their insight with the global tech community. You can make a general application, apply for a specific hot topic that we are recruiting an author for, or submit your own idea.

Hi!

I'm Morgan Evans, the author of Engineering Manager's Handbook. I really hope you enjoyed reading this book and found it useful for increasing your productivity and efficiency in the tools and knowledge necessary to thrive as an engineering manager.

It would really help me (and other potential readers!) if you could leave a review on Amazon sharing your thoughts on Engineering Manager's Handbook here.

Go to the link below or scan the QR code to leave your review:

https://packt.link/r/1803235357

Your review will help me to understand what's worked well in this book, and what could be improved upon for future editions, so it really is appreciated.

Best Wishes,

Morgan Evans

Download a free PDF copy of this book

Thanks for purchasing this book!

Do you like to read on the go but are unable to carry your print books everywhere?

Is your eBook purchase not compatible with the device of your choice?

Don't worry, now with every Packt book you get a DRM-free PDF version of that book at no cost.

Read anywhere, any place, on any device. Search, copy, and paste code from your favorite technical books directly into your application.

The perks don't stop there, you can get exclusive access to discounts, newsletters, and great free content in your inbox daily

Follow these simple steps to get the benefits:

1. Scan the QR code or visit the link below

https://packt.link/free-ebook/9781803235356

2. Submit your proof of purchase

3. That's it! We'll send your free PDF and other benefits to your email directly